Master Farmer

About the Book and Author

The story of postwar Greece holds invaluable lessons for many developing countries today. In 1947 Greece had just emerged from a decade of war and strife; its villagers were demoralized and fleeing rural life for the cities; and its farms were unable to produce adequate crops to feed its people. In less than forty years Greece has become a major exporter of foodstuffs, most villages have made the transition from underdeveloped to developing, and rural people no longer yearn to move to the cities.

In this book, Bruce Lansdale, director of the American Farm School in Thessaloniki, Greece—a hands-on educational institution for Greek villagers—provides an eloquent analysis of the training of so-called master farmers, who combine practical and theoretical knowledge of agriculture. Central to the Farm School's philosophy is faith in the capacity of peasants to solve their own problems and to accelerate the development process in agriculture. The school has concentrated on teaching both rural people and development workers the essential elements of management—planning, organization, leading, controlling, and adjusting.

In clear, concise terms, illustrated with delightful William Papas drawings of the legendary folk figure Nazredin Hodja, Mr. Lansdale outlines the philosophy and techniques of training master farmers. The book is addressed to development workers and those who train them and offers valuable insights into nonformal education, organizing short-course centers, managing secondary agricultural schools, operating student projects, and the problems of technology transfer.

Bruce Lansdale has spent thirty-seven years working in Greece for the American Farm School; he assumed the directorship in 1955. Under his guidance the school has continued its tradition of introducing innovative agricultural technology and supporting rural community development efforts.

Master Farmer

Teaching Small Farmers Management

Bruce M. Lansdale

Hodja drawings contributed by William Papas

LONDON AND NEW YORK

First published 1986 by Westview Press, Inc.

Published 2018 by Routledge
52 Vanderbilt Avenue, New York, NY 10017
2 Park Square, Milton Park, Abingdon, Oxon OX14 4RN

Routledge is an imprint of the Taylor & Francis Group, an informa business

Library of Congress Cataloging-in-Publication Data
Lansdale, Bruce M., 1925-
Master farmer.
Bibliography: p.
Includes index.
1. Farm management—Study and teaching—Greece.
2. Farms, Small—Management—Study and teaching—Greece. 3. Farmers—Training of—Greece. 4. Farm management—Study and teaching—Developing countries.
5. Farms, Small—Management—Study and teaching—Developing countries. 6. Farmers—Training of—Developing countries. I. Title.
S562.5.L35 1986 630'.68 86-5476

ISBN 13: 978-0-367-00634-1 (hbk)
ISBN 13: 978-0-367-15621-3 (pbk)

To Tad,
David, Jeff, Christine, and Michael
because we learned together
from our peasant friends

Contents

Preface

A lifetime of close association and shared experiences with Greek villagers has instilled in me a deep-seated faith in their ability to solve their problems, if they are given the opportunity and adequate support. These observations reinforce another of my convictions: Properly trained development workers can accelerate this development process through their leadership, dedication, and involvement in rural education as well as through their example as managers. Every bit of progress made by the peasants strengthens the confidence of the development workers, who in turn are able to be more helpful to the peasant. Each seems to contribute to the other's effectiveness, and they become part of a common process rather than working separately toward the same objective.

A key link in rural development has been the sergeants of agriculture: the specialized technicians who have both a practical and a theoretical understanding of farming and rural living. Some are farmers or farm workers; others serve the farmers or provide badly needed assistance to researchers, extension agents, and other professionals working with rural people. Some of these technicians have been trained in agricultural schools; others acquired their practical and theoretical expertise through years of experience supplemented by specialized courses in training centers and extension programs. This book was written to help development workers understand the practical philosophy and methodology of this training as it has been applied in Greece and other countries. It is addressed to those who want to help rural people in their own or other countries—agricultural extension agents, home economists, agricultural missionaries, development workers, and Peace Corps volunteers who are already in the field—and also to the large contingent of armchair strategists who have their own theories on development work.

To most people in Greece thirty-five years ago, the idea of teaching peasants to become managers would have seemed as incongruous as the suggestion today that village boys and girls be required to take a course in microcomputers. And yet the transformation of Greek agriculture and of the peasants has been so rapid in recent decades that management has become an essential element of peasant life, just as the computer will probably become their most indispensable tool by the turn of the century.

Many people with urban backgrounds, and even some from rural areas, think of peasants as second-class citizens, a view that is definitely not shared by this author. Derived from the Latin word *pagensis* (belonging to the country), *peasant* is defined in the *Encyclopaedia Britannica* as "countryman, either working for others or owning or renting and working by his own labour a small plot of ground."[1] In most societies the word *peasant* has disparaging connotations. In this book I have used the term with affection, admiration, and profound gratitude to my peasant friends for what they have taught me during the forty years I have worked with them.

Peasants played an important role in preserving the culture and values of Greece during the four hundred years of Turkish occupation, and they are the foundation on which any developing society must build its future. Even in countries and regions in which farmers represent fewer than ten percent of the inhabitants, such as in Western Europe, rural values continue to exert a salutary influence on the life of the country.

At the midpoint of this century the Greek villages had just emerged from a decade of war and civil strife. The people were hungry, poorly clothed, and demoralized. The majority of the peasants longed to leave the villages, and almost no parents wanted their children to stay. The villagers complained constantly about their misfortunes and were convinced that they could do nothing to overcome them. If asked who was at fault, they would blame the central government, the politicians, the civil servants, the military, the village president, or the "foreign factor," but they seldom blamed themselves. Fate had given them too many children and not enough or too much rain and had brought distress to their families, their animals, and their crops.

Many Greek and foreign development workers came to the villages with the idealistic conviction that they understood the peasants' problems and could solve them. However, they soon realized how important it was for the peasants to learn to identify and solve their own problems. Others could introduce modern techniques to the peasants and share new ideas with them, but until the peasants assimilated the ideas, they would have little meaning. The word *economics* is derived from the Greek word *oikos* (home). Although any rural development program must rely on effective planning at the national level, development workers must recognize that ultimately success depends on basic economics in the peasant home.

Postwar aid programs provided villagers with food, clothing, machinery, and housing. But the more villagers were given, the more they seemed to want and the less they seemed to appreciate the help. Free gifts bred dislike, distrust, and laziness. Eventually work programs were devised through which the villagers were able to earn the food, clothing, and supplies they needed by working on village projects, such as the construction of water systems or roads. The most difficult aspect of the assistance programs was to find ways of helping the peasants identify their needs, motivating them to fulfill them, and teaching them the required technical skills.

Many problems in the rural areas were beyond the peasants' ability to solve. Main roads connected only the larger towns. Few villages had electricity, and water supplies for drinking or irrigation were insufficient. Villages lacked satisfactory credit systems; cooperatives were not working effectively; and schools were inadequately staffed and badly maintained. An emerging agricultural extension service for disseminating new ideas to the villages had limited ties with research institutes and universities. Graduates did not want to work in the villages because they preferred to think of themselves as scientists.

The authorities in the provincial capitals had little decisionmaking power, and those at town and village levels had even less. Physical and psychological distances between the village and the capital were so great that when village authorities sought guidance from the central government they found that their questions were misunderstood or that they received irrelevant answers.

Complicated development issues had to be resolved by experts in the capital, who made long studies resulting in lengthy reports, many of which were never implemented. In fact, many of the infrastructure studies eventually provided the base on which villages could subsequently design their own projects.

In 1947 Greece was consuming a million and a half tons of wheat and producing only seven hundred thousand tons. Increased food production and an equitable system of distribution of food supplies were urgent concerns. Greece was looking to richer countries for expanded technical and financial assistance as well as for direct food aid.

The problems of Greece in the mid-twentieth century were not very different from those of developing countries today. The key solution to the plight of people in these countries is a sustained increase in their own food production. They also need a more equitable system of food distribution as well as ways to expand their exports from farming, fishing, and forestry. They must concentrate on increasing the yields per acre from existing farm land rather than on expanding the acreage of arable land. Technical and financial aid are essential to development in these countries. Other factors are new sources of water for irrigation and fertilizers and more efficient use of organic materials currently being wasted. Insect and pest control are equally important, especially in areas infested with tsetse flies.

Third World countries can follow the example of Greece, where the peasant families themselves implemented new approaches to development, often based on the advice of extension and home economics workers. They increased their knowledge of the technical aspects of agriculture and became more effective managers, and they solved their own problems rather than waiting indefinitely for others to do so. The seven hundred and fifty million farmers in the agricultural labor force of the developing countries must increase their own productivity so that they can provide food for themselves and the booming urban populations. They must learn more about irrigation and drainage, improved seeds, fertilizers, plant protection, animal husbandry, and farm machinery, and they must become better managers, master farmers able to plan more carefully, organize, lead, control, and adjust to changing circumstances.

The success of development programs in Greece makes the country an especially useful case study. Wheat production has more than tripled since World War II; the country has become self-sufficient in rice and sugar production and has developed into a major exporter of fruit and vegetables for the European market. The majority of rural people have seen their areas make the transitions from underdeveloped to developing to developed within their lifetimes. Villagers no longer dream of moving to the cities, and many of the urban immigrants now envy the country cousins whom they left behind. The extension agents, home economists, teachers, social workers, public health workers, Agricultural Bank employees, and their associates in the private sector have been the key agents of change in the development process. National and international organizations have also contributed to this transition.

But much of the credit must go to the villagers themselves, particularly to the women, whose role in many cases has been even greater than that of their husbands. These master farmers and their wives are the sergeants of agriculture: the select group of peasants who have acquired the managerial skills to lead the country to its present high level of agricultural production. It is essential that the developing nations of the world discover innovative approaches to creating such master farmers and the technicians who support them, by providing adequate management and technical training for subsistence peasants.

Throughout the world agricultural development has been based on education, research, and extension. In Greece, too, university training has prepared better equipped researchers and extension agents. Research provided new seeds, breeds, insecticides, pesticides, and equipment. Extension services carried the innovations to the farmers and their families. The achievements of these agencies, supplemented by work in the private sector, have contributed significantly to the development of Greek agriculture.

Since its founding more than eighty years ago, the American Farm School in Thessaloniki has been one of several organizations involved in agricultural development. Although its original purpose was to train village boys, a girls' school, operated in cooperation with the British Quakers, was added after World War II. The first short-course center in Greece was established at the Farm School in 1947, in cooperation with the Ministry of Agriculture. Today more than forty such ministry centers throughout the country train peasants to become master farmers. Through years of seeking new solutions to the problems facing the Greek peasant, the Farm School staff has formulated a number of basic principles, which have been incorporated into this book.

As a result of its long association with both the villages and those working with the rural population, the school has become more Greek than American. The number of American staff members represents less than three percent of the total, and they are regarded as being more Greek than American. Trustees resident in Greece play a far more active role in shaping policy than they did in the early years. Support from within Greece is now greater than contributions from abroad. It has become essentially a Greek school for village boys and girls and farm families.

Graduates of the Farm School and similar agricultural schools as well as former trainees from short courses are employed in a variety of occupations besides farming. They are assistants to extension agents, research technicians, milk inspectors, farm machinery sales personnel, garage operators, and irrigation specialists. Even the graduates not directly involved in agriculture continue to have strong links with farming and rural living.

An equally important training program, organized in Greece at about the same time as the first short-course center, was a one-year practical course to train home economists for the villages. The graduates of this program, working with the women, have had a tremendous impact on village development.

Visitors from many countries ask how the Farm School can best serve as a model for Third World nations. Study visits by school staff to developing countries in Latin America and Africa confirm that many principles that have grown out of the Farm School program could be applied there. A grant from the Rockefeller Foundation made it possible to investigate and confirm this theory in various institutions in Central America, the United States, and several Common Market countries. A report on observations growing out of these trips and an attempt to relate them to the Farm School's program provide the basis for this book. It has been divided into three major sections, which describe the school's underlying philosophy of development and the techniques that have been used to implement it. Part One covers the background of Greek rural development and the significance of practical agricultural education as a contributing factor. It also reviews the variety of programs that have been introduced at the Farm School, some of which are examples to be followed whereas others illustrate mistakes to be avoided. Part Two deals with the primary objective: training peasants and development workers to be managers. In Part Three the five basic objectives of a successful agricultural education program for training the sergeants for rural development are analyzed, and suggestions are made on how they can be attained.

The outstanding characteristic among progressive peasant families and development workers is their willingness to be stretched beyond the limits that they previously thought possible. The task of effective training programs is to continually motivate them to reach out even further. Nikos Kazantzakis expressed this concept in the opening lines of his *Report to Greco:*[2]

> I am a bow in your hands, Lord,
> stretch me lest I rot.
> Don't stretch me too hard, Lord,
> lest I break.
> Go ahead and stretch me, Lord,
> even if I break.

With adequate leadership they soon discover that they are far from reaching breaking point and possess almost unlimited potential. It is my hope that the extraordinary transformation achieved by the Greek peasant will inspire development workers and peasant families in Third World countries to seek new approaches to releasing the dormant potential in their own land.

Bruce M. Lansdale
Metamorphosis, Chalkidike, Greece

Acknowledgments

Two generations of American Farm School staff and both Greek and U.S. trustees have shown patience, understanding, and a willingness to share their wisdom with me since I first arrived there at the age of twenty-one. "Growing old I keep learning,"[1] Plato quoted Solon as saying, and I am grateful to those friends who have been my mentors for almost forty years.

A true teacher always finds time to offer insights and recognizes that his interruptions are his work. It is impossible for me to acknowledge individually the many institutional administrators, extension workers, home economists, teachers at all levels, and village men and women on five continents who have been so generous with their knowledge and their time.

One indispensable contributor to this book, who is believed to have been born between the thirteenth and fifteenth centuries, has so many names that he is difficult to identify. In Iran and much of the Arab world he is known as Mulla Nasredin, in Egypt as Goha, and in Turkey (where he is reputed to have been buried) as Nasredin Hodja. He is quoted by Muslims, Hindus, Christians, and Jews, often with the conviction that his tales are part of their own religious tradition.

Children delight in Hodja's amusing stories, adults are impressed by his wisdom, teachers are interested in his philosophy, and theologians

One day an illiterate neighbor brought Hodja a badly scribbled letter to read for him. When Hodja complained that it was illegible the man accused him of being unworthy of the "turban of wisdom" that he wore. Hodja was furious at the insult and slammed the turban on his neighbor's head. "Here," he shouted, "you wear it and see if you can read the letter!"

seek to understand his mysticism. His tales have taught me to laugh at myself in the most frustrating situtions—a vital lesson for any development worker. But reading Hodja stories is a poor substitute for listening to them beside the hearth in a peasant's hut or exchanging them for hours in a village coffee shop: experiences that are slowly disappearing in Greece along with the vanishing peasant. It is my hope that a sampling of his tales will arouse the reader's interest in learning more about him. To that end, I have included a listing of interesting books about Hodja at the end of the Notes section.

The drawings by Bill Papas of Hodja and Greek peasants who have been so much a part of his life express the love and compassion that they deserve. I would like to thank him for illustrating this book in a way that recreates peasants in their natural environment far more effectively than can the written word.

Nasredin Hodja, peasants, and teachers around the world inspired a myriad of disjointed ideas that are the basis of this book. I am particularly grateful to Marguerite Langford, who has helped me organize them in a form that I hope does not require Hodja's turban to be understood.

B.M.L.

PART ONE

Managing the Training

Training peasants to be well-rounded managers is one of the most difficult aspects of the development process. Greece's success in this area should provide inspiration to those attempting to teach management in rural development programs in the Third World, and the means by which development workers in Greece have achieved this success should prove informative. In a Greek agricultural community, the *nikokiris* (nee-ko-kēē-rees) is the capable manager of the farm; his wife, the *nikokira* (nee-ko-kee-rā), is in charge of the household and often the farm accounts. Neither term, when used to describe the peasant as master farmer, has a close equivalent in English.

Correspondingly, the word *management* has no equivalent in Greek. It is variously translated as administration, direction, and organization. Management consultants and others are thus obliged to use a transliteration of the English, pronounced "manatzment." Ita Hartnett, supervisor of farm and home management in County Cork, Ireland, defined management among rural families as the skill of "doing what you want with what you've got"—a phrase easily understood by peasants. In more sophisticated terms, the process can be described as "balancing objectives with available resources."

A network of well-managed agricultural schools and training centers has provided the backbone for development in many countries of Europe and North America. The graduates of these institutions have played a key role in the growth of agriculture and the improvement of rural living in their countries. The influence on development of the folk high schools in Denmark, agricultural institutes in Holland, and Future Farmers of America chapters in U.S. high schools has been enormous.

In developing countries, however, rural training institutions have been viewed with considerable skepticism. Coursework tends to be theoretical rather than practical, primarily because instructors lack hands-on experience. The cost per student is considered disproportionately high, and the required investment in land and buildings, too great. Graduates of these programs are eager to use their training in urban situations rather than return to their farms or seek jobs in occupations related to agriculture. Failure to clarify objectives results in poorly organized instruction; thus the graduates from these institutions are often inadequately prepared.

Although there is considerable validity in these criticisms of agricultural schools, often the problem lies in the failure of school administrators to organize and operate their institutions effectively. Politicians who are impatient to see immediate returns from their investments in education have discovered that they receive more direct return from expenditures in extension work and community development than from long-term investments required for training farm youth. Yet there is an urgent need for technicians in agriculture who have had both practical and theoretical training in either secondary agricultural schools or rural training centers. If developing countries are to make maximum use of their potential, doing what they want with what they have, they must find ways to distribute available resources to extension and community development *and* to the institutions that train the leaders needed in these programs.

In Part One of the book, the problems of training farmers in developing countries are explored, along with the general educational principles and specific contributions that can be made by agricultural schools and training centers.

1

Training Master Farmers

- What kind of training do peasants need?
- What should be the goals of agricultural education in a developing society?
- What are the shortcomings of educational systems in developing nations?

One day Hodja was sitting in front of his hut searching in the dust when a neighbor came by and asked him what he was looking for. When Hodja told him he had lost his gold coin, the neighbor kneeled down and started sifting through the dirt. After a while, when he asked the Hodja exactly where he had lost it, Hodja told him that it had disappeared in the hut. When the surprised neighbor asked why he was searching outside, Hodja replied, "It's much too dark inside the hut to look for it."

Peasants throughout the world have demonstrated their capacity to become master farmers—managers capable of planning, organizing, supervising, and overcoming the variety of challenges inherent in farm work and rural living—even though many development workers continue to look upon them as passive objects in technical assistance programs. Economists and other experts have tended to consult officials in the cities for solutions to the peasants' problems, bypassing the peasants, who they assume are incapable of resolving their own problems. Like Hodja, the specialists think it is "too dark inside." Perceptive village workers, however, quickly realize that peasants are eager to improve their way of life and have the potential to do so. Yet development agencies have generally relied on extension services and community development programs to supply the required help rather than providing institutional training programs for the peasants, which they have regarded with skepticism.

In the same way that formal and nonformal secondary and postsecondary vocational agricultural instruction has played a significant role in training master farmers in the West, specialized training is needed in Third World countries, even though the costs may be relatively high. Agricultural training is not a matter of selecting one method of training over another, but rather of developing an approach that integrates a variety of complementary educational techniques. Institutional training is emphasized throughout this book because this most important element of development has been generally neglected.

UNDERSTANDING THE PROBLEM

The outlook of Greek peasants during the early part of this century was similar to that of peasants in Third World countries today. The peasants generally accepted their feelings of helplessness in the face of constant misfortune as inevitable and played a primary role in initiating it. Their misfortune was something neither they nor those who wanted to help them could do much about. In Greece today, progressive farmers have grown to accept change as inevitable. If development workers in Greece and elsewhere are to utilize the peasants' potential for change, they must first understand the characteristics of the villagers that once hindered progress in the Greek countryside.

1. The peasants lacked self-confidence. Few considered themselves to be "winners." In fact the term *peasant* was almost synonymous with *loser.* They took for granted that anyone with ability or ambition would leave the village, either to be educated or to earn a better living. They assumed that education was unnecessary for the less talented and for women, whose functions were to bear children and care for the home.

Those lacking drive were expected to remain in the village and cultivate the family land.

2. The peasants had had no management training; they did not understand how to plan or organize their operations or how to maintain adequate records. Some shepherds, particularly the leaders of their clans, had learned rudimentary management from their fathers or by trial and error, and a few farmers were shrewd businessmen; but these were the exceptions. Because of custom, the many resourceful women were unable to express themselves except in very subtle ways, so their management abilities were only marginally utilized.

3. Villagers were unwilling to accept advice from outsiders. As most innovative approaches ended in failure and they lived on such limited budgets, the villagers could not afford to take chances. In 1946 few Greek peasants grew rice and none believed that the crop could be produced on a large scale; yet today Greece is an exporter of rice.

4. In addition to being ill equipped to solve such problems as pest control for crops and disease among animals, peasants were weak in problem-solving skills and were thus at a disadvantage when dealing with shrewd merchants. If their field lacked a basic fertilizer or trace element, they found it easier to apply a balanced nitrogen-phosphate-potash mix than to find out what was missing. If their chickens did not lay eggs, they blamed either the evil eye or the chickens but seldom considered adding protein to the feed.

5. The agricultural knowledge of the peasants was restricted to current village practices. Rural schools taught little or nothing about modern agriculture or village development. Medical knowledge was limited to remedies that had been passed down through generations of villagers. The arithmetic they had studied in local schools was unrelated to such practical calculations as the area of their fields and the percentages of chemicals in fertilizers. Their practical knowledge was limited to their own experience, and they showed little interest in acquiring more.

6. The peasants lacked manual skills and dexterity, and they had to hire technicians for most building or maintenance work. Only in recent years have they recognized that they themselves can do elementary carpentry, masonry, plumbing, or electrical repairs and perform routine maintenance on their buildings and equipment.

THEORY AND PRACTICE

Essential to the progress of rural development is an educational system based on indigenous needs rather than a system transplanted from an alien culture. In the first half of this century Greece's developing society stressed the traditional educational system, which was not designed to meet the requirements of educating both skilled master farmers and theoretically trained bureaucrats. Likewise the Greek culture tended to emphasize the value of theoretical knowledge at the expense of practical experience, leading those peasants who succeeded in the traditional system to pursue another life-style than village farming. A development worker can see in the following description of Greek education instances at every institutional level in which the schools did not meet the needs of the peasant farmer. This continues to be one of the major problems of education in the Third World today.

At the turn of the century, the city dwellers' stereotype of the Greek peasant was a manual worker who had limited theoretical knowledge. They failed to appreciate his close relationship with the natural world around him or to recognize his wisdom. They assumed that his limited knowledge of urban life and thought indicated a corresponding lack of intelligence.

Traditional Education

The first step toward development in Greece during the four hundred years of Turkish occupation was the organization of elementary schools by village priests to teach Greek children reading and writing. However, as the children of peasants became more educated, they were less inclined to do manual work. This generation of young people was thankful to be relieved of the drudgery of farming and to be able to devote more time to their studies than to helping their parents with their work.

After 1821 when Southern Greece was liberated from the Ottoman Empire, the government introduced the high school, modeled after the French lycée and the German gymnasium. Here children were educated entirely by rote learning, with emphasis on classical Greek. The student who obtained high grades in examinations was much admired. Academic education became a way to escape the manual work of the farm. The newly educated young people developed a certain smugness and sense of superiority, especially toward their parents. Most of them rejected farming and village life in favor of the attractions of urban living.

Finally, at the university level, schools of agriculture were established to prepare agriculturists to work with the farmers. Many of the students at these schools came from the cities or had spent their high school years there and had had little practical training. Because of the absence of practical experience in farming and the premium placed by these students on theoretical knowledge, the university trainee often was unskilled in the very areas in which he was to instruct the villagers. He was not versed in troubleshooting with machinery, had little practical experience in caring for livestock, was often unable to diagnose or treat diseases in plants and animals, and lacked self-confidence, especially in his early years after graduation. However, because of his university education, he felt superior to the lowly, uneducated peasant.

More Than a Diploma

Training master farmers requires different programs from the traditional educational format exported by Western countries to developing nations. "Diploma-itis" is one of the worst plagues ever foisted on the Third World. Master farmers need more than the diploma obtained by memorizing a vast body of facts: They must have an educational system that stimulates them to use their minds to solve problems, they must work more productively with their hands, and they must acquire greater self-confidence. A Greek student discussing his five years at the university agricultural school was surprised when someone expressed an interest in what he had learned to do. It was not the education that counted for him but rather the "five spot"—the passing grade that allowed him to receive a diploma qualifying him for the civil service.

President Julius K. Nyerere of Tanzania described the need for nontraditional education in his classic article, "Education for Self-Reliance": "The education provided must therefore encourage the development in each citizen of three things: an inquiring mind; an ability to learn from what others do, and reject or adapt it to his own needs; and a basic confidence in his own position as a free and equal member of the society, who values others and is valued by them for what he does and not for what he obtains."[1]

This was the approach to education that Dr. John Henry House felt was needed in the Balkans twenty years before he founded the American Farm School at the beginning of the twentieth century:

> Early in my taking up educational work I discovered what seemed to me to be a great defect in our educational system. I found that boys who had a secondary and high school education were supposed not to dirty their hands with manual labor. Village boys were being educated away from their villages and were being taught that they had entered into a higher social class. Labor with the hands was considered beneath them. I said we are giving these boys a wrong education. My mind immediately started to work out a plan of education which should train children along the heart, the head and the hands. It was years afterwards that I was able to see my ideal realized in the Thessalonica Agricultural and Industrial Institute.[2]

Even today, however, planners seem to have difficulty grasping the implications of Nyerere's and House's ideas and expressing them to the students.

SERGEANTS OF AGRICULTURE

Another way to understand the problem of training master farmers is to look at the structure of agricultural development in military terms. A well-organized army consists of three levels of personnel: privates, noncommissioned officers, and commissioned officers. The privates are the army's foundation, whereas the commissioned officers, graduates of officer training schools, provide leadership and planning. However, the key to an army's successful operation lies with the sergeants, whose basic training and practical experience enable them to understand every detail of how the army works.

In Greece many peasants and farm workers who were accustomed to long hours of back-breaking work had much in common with privates in the army. They were content to cultivate their crops and raise their animals without feeling any strong desire to improve their lot, just as some privates are content to remain as they are without wanting to be promoted to positions of greater responsibility. In every village there is a corps of peasants who are satisfied with their way of life and lack any strong ambition to change it.

Even in recent years the many university-trained commissioned officers in agriculture have a theoretical understanding of farming but limited practical experience until they have worked for several years following graduation. They have very little contact with the day-to-day problems of farming and lack the ability to relate to the farmer. One group of university-trained agriculturists who had considerable practical farm experience referred to those who grew up in cities and worked in the Ministry of Agriculture immediately following graduation as "asphalt agriculturists."

A real problem in Greek agricultural development has been the shortage of agricultural sergeants: those who link the peasant and the agriculturist and who are trained and equipped to deal with the practical aspects of daily farming. It is as important to provide specialized secondary education for farmers so that they can become the capable managers so urgently needed in developing countries as it is to train sergeants in the army.

The military sector uses two types of sergeants: those with line functions and those with staff responsibilities. Line sergeants—the leaders of men—are responsible for running a platoon and seeing that the orders from their superiors are carried out. Because they have been trained and are confident in what they are doing, they command respect from the men who follow them. But a good army also needs staff or technical sergeants, who provide servicing and specialized skills. They have the knowledge, experience, and practical ability to deal with almost any problem within their sphere of competence.

Sergeants in the army are the product of two different forms of training, which should also be available to the sergeants of agriculture. In the army those who have come up through the ranks over a period of years are from the school of hard knocks. They have been promoted periodically until they have reached a position of authority and command. Similarly, some farmers in each village have become competent by learning from their own mistakes and those of others. Because they have had to make choices, they have become effective decision makers and know that no problem has only a single solution. The difficulty in a rapidly developing society is that there are not enough sergeants of agriculture with sufficient training in innovation and problem solving to deal effectively with the ever-changing challenges that farmers face.

Because a private would need to serve for many years in the regular army to acquire the qualities of a competent sergeant, the military has organized a second form of training in special schools to shorten this training period. Corresponding schools are needed in the developing countries to train sergeants for agriculture; unfortunately there are few of these. Just as the army instructs its personnel in management, so these schools must train farm youth for their role as master farmers or agricultural technicians.

The qualities of an army sergeant provide an interesting guideline for the requirements of the ideal master farmer. Like his counterpart in the army he must be motivated to rise above the level of the private. He must emulate the essential attributes of an effective sergeant so that he can lead as well as follow. Technical competence growing out of personal experience and organizational skills are basic requirements. He should have confidence in himself and be open to the suggestions of others. Flexibility is important to him in dealing with changing conditions. He must have the necessary basic knowledge and ability to solve the various problems that constantly arise. Like the good sergeant the master farmer must be an able manager, which in essence is a combination of all these characteristics.

KAKOMIRIS AND NIKOKIRIS

Two Greek words, which have no real equivalents in English, *kakomiris* and *nikokiris*, can be used to distinguish two extreme types of agriculturalists: those who are satisfied with their lot (privates in agriculture) and the master farmers (the sergeants). The kakomiris (literally translated as "the ill-fated one") is the peasant whose luck is down or who has suffered some natural misfortune. Discussion with many villagers confirms that the kakomiris almost always brings the bad luck on himself through lack of adequate planning, even though he is inclined to blame others for all his setbacks. A frequent visitor to the Farm School often describes himself as a kakomiris. Whenever something goes wrong he can be heard heaving a great sigh and saying, "Ah, o kakomiris, o Giorgos" (poor ill-fated George), reinforcing his conviction that he is a born loser.

In direct contrast to the kakomiris is the nikokiris (literally translated as "the master of the house").[3] He initiates action, plans and organizes his work effectively, employs other peasants, and takes advantage of every opportunity to increase his income and the size of his holdings. The nikokiris is a leader in the village, respected by others for his judgment. Villagers turn to him for support in times of crisis. The disparity between the appearances of the homes, fields, and machinery of the nikokiris and those of the kakomiris are readily apparent. The terms arrayed in Table 1 can be used to distinguish the typical thought and behavior patterns of the kakomiris and nikokiris in a variety of situations. Although the examples may be extreme, they do provide a basis for comparing the way in which each sees himself and those around him.

It would be helpful to identify terms equivalent to kakomiris and nikokiris in the languages of Third World countries to determine whether the characterizations are valid in developing societies elsewhere. Farmers with equivalent traits can be found even in the more advanced nations of Western Europe and North America. Just as industrial managers have discovered that it is profitable to train unskilled workers to become master technicians, it is also important that corresponding training be made available to the rural people in developing societies. Institutional training programs must help peasants and farm workers feel more positive about themselves and acquire the knowledge, the competencies, and the attitudes of master farmers in order to become effective managers of their enterprises, regardless of size or scope.

The ever-increasing proportion of nikokiris-type peasants has played a significant role in transforming the Greek villages from backward, dormant communities into dynamic, progressive societies within this century. These farmers are capable managers, better able to plan their lives and organize their work. They have acquired the knowledge and skills to deal more effectively with their problems. These are the men and women who have led Greece to new heights of agricultural production while creating a more satisfying way of life for themselves.

TABLE 1. Thought and behavior patterns of kakomiris and nikokiris

	Kakomiris	Nikokiris
Knowledge		
Admitting ignorance	Feels he knows it all	Admits he doesn't know everything
Acquiring new knowledge	Feels he knows enough	Continually learning
Sources of information	Never reads, seldom inquires	Consults others, reads periodicals, asks questions
Competencies		
Craftmanship	Careless, no attention to detail	Neat and thorough
Learning skills	Casual, inattentive	Learns carefully
Using skills	Slovenly, indifferent	Enjoys working as craftsman
Attitudes		
Toward others	Suspicious, fearful	Cautious but confident
Toward new ideas	Resistant to change	Willing to try
Toward family	Strained, critical	Relaxed, supportive
Toward children	Critical, fault finding	Responsive, warm
Toward neighbors	Strained, unfriendly	Helpful, friendly
Problem-solving techniques		
Problem-solving approach	Jumps to conclusions	Analyzes carefully
Problem-solving orientation	People oriented, subjective	Problem oriented, objective
Problem-solving ability	Ineffective	Capable problem solver
Using outside help	Tries to go it alone	Seeks advice
Cultural barriers	Limited by culture	Willing to seek new solutions
Self-esteem		
Feelings about self	Very low, loser	Sense of confidence, winner
Feelings about family	Tends to deprecate	Proud of family
Encouragement of children	Puts them down	Builds them up
Place in society	Low status	Leadership
Attitude of others to him	Pitied	Respected
Emotional stability	Volatile	Steady, dependable
Management abilities		
Confidence	Lacks self-confidence	Strong self-confidence
"What he wants with what he's got"	Spends beyond means	Lives within means
Responsibility	Blames others or fate	Accepts responsibility for mistakes
Business enterprise	Out of control	In full command

TABLE 1. (cont.)

	Kakomiris	Nikokiris
Planning		
Short-term planning	Depends on circumstances	Plans daily activities
Medium-term planning	Waits to see	Anticipates problems
Long-term planning	No clear objectives	Clear-cut, long-term goals
Use of time	In coffee shop, visiting	At work
His fields as capital	Sells fields	Buys fields
Organizing		
His home	Dilapidated	Well kept, neat
His farming	Barely subsistence level	Businessman farmer
His installations	Disintegrating	Maintains old, building new
Capital investment	Consumes capital	Increases capital
His machinery	Rusted, badly maintained	Painted, well kept
General organization	Poorly organized	Well organized
Leadership qualities		
Attitude toward work	Apathetic	Energetic, enthusiastic
Relation to workers	Indifferent	Inspires confidence
Cooperation with others	Tends to be loner	Cooperates well
Supervision	Negative	Positive
Treatment of livestock	Treats poorly	Loving care
Control		
Record keeping	No records	Careful records
Farm inventory	No inventory	Detailed inventory
Appearance of farm	Sloppy	Well kept
Adjustment		
Flexibility	Rigid in management	Adjusts plans to circumstances
Consultation with others	Goes own fixed way	Works with others

A neighbor borrowed Hodja's donkey so often that one day he decided to put a stop to it by telling him the donkey was dead. He locked it in the barn and sat on his front steps to await the neighbor. As he approached, Hodja burst into crocodile tears over the death of their beloved animal. When the neighbor heard the news he sat down next to Hodja, lamenting the loss of the donkey and remembering how useful it had been. They were both surprised when at that moment the donkey began to bray. Infuriated, the neighbor turned to Hodja and accused him of being a liar. "What," said Hodja, "you believe my donkey and you don't believe me!"

2

A Society in Transition

- Why is the Greek village significant as a model for development?
- Are there similarities between the attitudes of rural people in Greece and those in Third World countries?
- How do peasant attitudes relate to development?

Most expatriates who have lived in rural Greece for years find it difficult to be objective when speaking about their peasant friends because of their personal affection, admiration, and close identification with the people. They particularly appreciate the stamina that the villagers have shown for centuries under adverse circumstances, their simple and genuine hospitality, their delightful sense of humor, and the joy they find in so many basic things. Newcomers are often amused by the apparently biased enthusiasm of "old Greek hands" and, like Hodja's neighbor, never quite believe what they hear.

The attitudes of the more sympathetic city people and foreigners toward the Greek peasant after World War II expressed the paradox of development. They urged him to progress and at the same time expected him to be changeless. They cherished their leisure with him but encouraged him to be more active and organized. They enjoyed the primitiveness of his village but were forever telling him to build a better house and dairy barn and keep his cows cleaner. They wanted to sing and dance with him, yet insisted that he should make better use of his time.

A chorus of criticism by less understanding urban Greeks and expatriates during the first half of the century is typical of comments heard in the Third World today. "Why are they so lazy?" "I find them pigheaded." "Their resistance to change is so frustrating." "What makes them so suspicious?" "They will always be the same." "They are incapable of managing their own lives." "The only way to deal with them is with strong authority."

It is helpful to trace the transition of the Greek village during this century and to try to identify elements in the process that might be useful for understanding problems of development in the Third World. The most obvious changes have been in external manifestations—homes and public buildings, personal appearance and clothing, health and sanitation, mechanization in agriculture, and the impact of rural electrification. But the more subtle differences in the way people think, their attitudes about themselves and others, and their ability to deal with complex problems indicate the fundamental changes that have taken place.

AT THE TURN OF THE CENTURY

It may be difficult to appreciate that parts of Greece—which has a tradition of culture and philosophy that goes back twenty-five hundred years—were occupied territories within the lifetime of its recent president, Constantine Karamanlis. At the time of his birth just after the turn of the century, his village and all of Northern Greece had been under

Turkish occupation for four hundred years: four centuries of stagnation and decline. Most major public buildings throughout Greece date from either the classical or Byzantine periods before the Turks overran the country or the nineteenth or twentieth centuries; very few buildings of architectural significance were constructed between these times.

One way to help those unfamiliar with Greece understand what the country was like seventy-five years ago is to describe the area around Karamanlis' village of Proti in Northern Greece. At that time the village was known by the Turkish name, Kiupkioi—town of the pots.

In villages a few miles to the west, the Strymon River, source of both hope and fear for the people, overflowed its banks every few years, leaving destruction and poverty in its wake. Malaria, typhoid fever, and tuberculosis were rampant. The life expectancy of the local people was less than forty years.[1] Most families had six to eight children, hoping that at least half might survive. The birthrate was more than twenty-five per thousand as compared with fifteen today, and the deathrate was more than 15 per thousand as compared to fewer than nine today.[2] Lice and fleas were commonplace so that most children had their heads shaved. Visitors returning from distant villages were sent to the laundry to bathe before entering any respectable home. The per capita income is estimated to have been less than $100 per year.

Farm power came from donkeys, oxen pulling wooden plows that had changed little in two thousand years, and human labor, especially that of women, who had to wield mattocks to hoe their fields. It required one hundred and twenty-six worker hours to produce a hundred kilograms of wheat compared to one and one-fourth worker hours in Greece today and far less in the United States.[3]

In some villages people lived in simple, whitewashed, two-roomed stone or mud-brick houses with thatched or mud-tiled roofs. Families slept on straw mats on the earthen floor, which they covered with special mud every month to keep down the dust. In tobacco villages like Proti, families lived on the second floor of two-story houses and used the lower floor to house animals and store crops. Their diet was primarily bread, cheese, olives, and vegetables in season, with garlic to ward off disease. Milk was for sick people and babies. Meat was eaten on feast days, and chicken was considered a great luxury. This was not Ethiopia, Tanzania, or one of the poorer countries of Central America, but a Greek village at the beginning of the twentieth century.

The middle of the twentieth century following World War II and the Greek Civil War was the turning point in Greek village life; it changed from simple rural existence based predominantly on subsistence farming to an agricultural society in transition. Who were the peasants at that time? What were they like? How did they think? Foreigners

Greek Villagers in the Mid-Twentieth Century

The peasants are extremely honest and yet they distrust one another. They are truly humble but intensely proud. They are devotedly loyal to their employer but can be devastatingly critical. Most Greeks are convinced that they could do as well as the Prime Minister in running the country if only they had the chance.

The villagers are deeply religious yet they castigate their church. When a priest walks by, peasants can occasionally be seen tying a knot in their handkerchief. They will tell you that the priest carries the evil eye because the devil walks three feet behind him, and how else could they protect themselves but by tying a knot in their handkerchief?

They are the essence of conservatism, yet they do not hesitate to try the radical. The men are devoted to their family, yet they believe freely in a double standard and never seem to miss an opportunity. This privilege has always been accorded to men but never to women, who are expected to be loyal to their husbands. The men play the dominant role, yet the women run them.

Villages are tremendously hospitable, yet they are often suspicious of a stranger. They are completely united in time of war and just as completely divided in peacetime. They are dedicated to their past yet are pragmatic about their present. They spend hours criticizing their politicians yet they immediately seek help from their member of parliament whenever they have a problem.

found them a fascinating, paradoxical people, difficult to understand and even harder to describe. Some impressions written more than thirty years ago based on the author's observations are helpful in understanding the complexity of peasant character at that time (see above).

CHARACTERISTICS OF GREEK PEASANTS

Two characteristics of Greek peasants, which baffled foreigners, are described by the words *philotimo* and *philoxenia*. In the Greek language, *philotimo* originates from two words, *philo* (love for) and *timi* (honor), and *philoxenia*, from *philo* (love for) and *xenos* (stranger). The first trait is an amalgam of pride, ambition, ego, face, honor, and dignity. It incorporates the peasants' sense of status, importance of their family and its origins, the village from which they come, and loyalty to their country. Those who understood this quality found it to be the villagers' greatest virtue; those who did not, regarded it as their worst fault. If someone appealed to this aspect, there was nothing in the world the peasant would not do for him, but if he tread on it, there was nothing the peasant would do for him. Philotimo was very easy to tread on and very difficult to appeal to.

Philoxenia, unstinting hospitality to the stranger, is a tradition that dates back to ancient Greece, a time when a person was never certain whether a visitor was a god in disguise. It has always been delightful to experience this warmth, especially in a distant village on a cold, rainy night. However, many development workers failed to realize that hospitality had its limits; once they settled in a village they had to be careful not to take it for granted. Reciprocation brought a new and far richer relationship. If a Greek peasant saw that a stranger needed help, he would assist him in any way he could, whether in the remotest village or the heart of a city.

Development workers found it difficult to understand some of the subtle personality traits of the villagers that might hinder development. The villagers often expressed themselves by using stereotyped phrases accompanied by characteristic gestures.[4] A group of Greek and expatriate associates who had lived in the villages attempted to list these phrases with their implicit meanings (see Table 2). Anyone working in development should look for similar expressions and their meanings in other countries; they provide a very useful insight into the culture.

When a peasant said, "It doesn't matter" or "Who cares?" he would shrug his shoulders and stick out his hands, palms upward, with an unconcerned look on his face, indicating that he was indifferent to seemingly irrelevant detail. Villagers occasionally took this attitude about changing the oil in their tractors when farm machinery was first introduced in Greece. Their response was often, "It doesn't matter," until they discovered a burned-out bearing. At that point they blamed the man who sold the tractor, the manufacturer, or the tractor itself for the damage. Or they used the phrase, "That dishonest thing"—blaming the tractor for having broken down.

If the peasant nodded his head in a positive manner and used the phrase "I understand" when an agriculturist was trying to explain something new to him, he really meant "Don't bother to tell me. I understand what you are saying before you open your mouth." He was both anxious to avoid appearing ignorant and eager to show how quickly his mind worked. In contrast, the phrase "It can't be done" was always accompanied by raised eyebrows and hands placed forward with the palms upright. These gestures expressed not only doubt but determination that something that had not previously been accomplished could not

TABLE 2. Phrases used by peasants in midcentury Greece with their literal and implied meanings

Greek Phrase	Literal Translation	Implied Meaning
Then ginete	It can't be done.	It hasn't been done before so it can't be done.
Katalava, katalava	I understand, I understand.	I've got a much faster mind than you think.
Opou nanai erhetai	Any minute now.	Wherever he is (and I don't really know) he will be coming (and I'm not sure when).
Then pirazi	It doesn't matter.	Don't bother me with details — they don't make much difference, and besides, who cares?
To atimo	The dishonest thing.	It's the thing's fault for breaking down, not mine.
Na to skefto	Let me think about it.	Give me a chance to ask my wife what she thinks.
Xeries pios eime ego	You know who I am?	I'm as good as you are 'cause I'm related to a VIP and don't you forget it.
Tha ta volepsoume	We'll take care of it.	I haven't really figured it out, but don't you worry about it — I'll find a solution.
Tha se kanoniso	I'll fix you.	I'll get even with you if it's the last thing I do.
Then ftaio ego	It's not my fault.	Don't blame me. It's someone else's fault and I can prove it.

be done, even if the villager had to sabotage the effort to prove his point.

Such phrases as "I'll take care of it" or "I'll fix it" usually expressed the peasant's optimism. As he said it he waved one hand in the air as if there were no problem, while appearing to be in complete control of the situation. In fact, it usually implied that he had not thought out how he would resolve the difficulty but was confident that he could use his wit to find a solution.

"It's not my fault" was the thought behind many actions and decisions of civil servants in the villages and by the villagers as well. They seemed to plan activities so as to avoid any responsibility should things go wrong. But when something did not turn out as hoped, a chorus of voices from villagers with hands and eyebrows upraised would respond, "It's not my fault." Few would say, "Let's see what went wrong and find a solution."

The phrase "You know who I am?" was related to a villager's sense of pride. He may only have been a distant cousin of a member of parliament, but the connection made him feel significant. Nothing infuriated him more than an outsider's failure to recognize his sense of importance. A look of deep distress would appear on his face, making the outsider wonder if he would ever be forgiven.

The Greek equivalent of "in a little while" literally translated meant, "wherever he may be he will be coming." If a villager was asked when someone was coming and he responded with this phrase and an eager look, he meant "He is somewhere, I don't know where, but wherever he is he is probably on his way." Development workers always felt deeply frustrated until they understood that "in a minute" might be any time during the next few days.

PATTERNS OF BEHAVIOR AMONG PEASANTS

Even though the attitudes, work patterns, and relationships of peasants do evolve, an elderly Athenian businessman maintains that it requires three generations for basic character traits to change significantly. He insists that this is particularly true in rural areas, although even in the cities, where the newly arrived villagers dress smartly and live in large apartment houses, he is convinced that their way of thinking has changed little.

As an indirect consequence of years of foreign occupation, villagers always had a deep-seated distrust of anyone outside the immediate family, especially city people and officials. In the village, only members of the family, encompassing second cousins or godparents, were trusted; however, when peasants were far from home, they considered all fellow villagers members of their family. The closeness and the interdependence among family members caused villagers to place a very high premium on children. This value was touchingly expressed by a white-haired shepherd, who commented to a visiting foreigner as he fondly embraced his grandchildren, "The real gold in life is made of children."

Peasants often had misgivings about the motives of civil servants. They suspected that all government employees wanted to exploit the villagers and take advantage of their positions, while doing a minimum of work. Some of the finest people in Greece are civil servants, but unfortunately the villagers had enough experience with apathetic government workers to justify their fears.

Low mobility and lack of communication with other villages were widespread in peasant societies. As recently as 1958 a villager who lived less than one hundred kilometers from the Farm School could not be persuaded to travel this distance for a short course. Now with roads, a frequent bus service, and private cars this same person thinks nothing of driving a hundred kilometers and back to the city just to shop.

Living in a central village with surrounding fields established a clear sense of community among the peasants. They felt that they belonged, recognized their leaders, and knew every detail of each person's activity within their village. In the growing urban centers the absence among young people of a sense of identity with the community has become a major threat in contemporary Greek society. An Athens taxi driver, describing this problem, related it to the noxious exhaust fumes. He pointed out that not only what the new arrivals smelled but what they saw, touched, and heard affected them adversely. He added, rather ruefully, "But the worst part of it is that after a while it's what we say and do that stinks." Perhaps this lack of a sense of belonging, of the embrace of the village, was the peasant's greatest loss when he moved to the city.

For many years men in the village enjoyed their leisure time without feeling guilty. They were accustomed to a life-style that included underemployment, seasonal work, and the frustrations inherent in a developing society. They realized that much should be done, but they felt they lacked the skills and the materials to do it. They developed a tendency to procrastinate, giving them time to sit in the coffee shop. Although custom required the women to appear busy, they found the leisure to exchange local gossip when they gathered to collect water at the village well. Even here they often carried their spinning or knitting.

To outsiders the Greek peasants may have appeared lazy. Visitors who shared their back-breaking, eighteen-hour day during the two-month tobacco harvest realized how false this impression was. Peasants were wise enough to know when it paid to work and when to relax and enjoy life.

One reason why villagers hesitated to accept new ideas was that they found so much of their life-style satisfying and did not want to change it.[5] A Macedonian shepherd with more than a thousand sheep was advised by his son, a Farm School graduate, to replace them with cows. The old man told a visitor sadly that his son might be right, but if he did not have the sheep to wake up for before dawn, to milk, and to renew his life each spring, he would wither away. "After I die he can sell them all," he said, "but until then I told him to keep three hundred for me to look after.

The Greek peasant was deeply superstitious. If asked about the evil eye he would at first deny its existence, but after some encouragement he would admit that it was a dynamic force in his life. This belief in the supernatural was expressed by a woman who had just broken her arm and, assisted by her husband, was walking to the nearest town to find a doctor. They told a man who gave them a lift that a neighbor had put the evil eye on her, assuring him that everyone in the village knew that this woman cast the evil eye. They also said that the priest and certain women with particular skills knew how to remove it. Related to this belief was the villagers' deep sense of fatalism.

Despite their arduous daily lives, peasant families have always shared a child-like joy. Even thirty-five years ago while suffering from the impact of the war years, they were able to find pleasure in fleeting moments of good fortune. As the peasant families have grown prosperous, they often recall the "good old days" with great nostalgia and speak of how much richer their lives were then. Remembering this period, a waiter who lived in a modern apartment building in Athens with a university professor, an army general, and several other people commented that the fellow tenants never greeted each other. "Back in the village at harvest time we used to carry our grapes home, offering a bunch to every friend along the way. Here in Athens we can't even offer our good morning to our neighbors."

THE GREEK VILLAGE TODAY

A dramatic change has taken place in the villages over the past thirty-five years. Anyone visiting former President Karamanlis' birthplace, Proti, today would find it difficult to believe that it is the same village described at the beginning of this chapter. Most of the homes have television sets, comfortable furniture, and souvenirs of trips to Athens or even neighboring countries. Many of the residents earned enough money working in Germany to buy a pickup truck or a refrigerator, or even to build a new house. Low-interest credit from the government or the Agricultural Bank has helped others to renovate their homes or build new ones.

A paved road leads into the village from the district capital, and many of the internal roads are now surfaced. Malaria and typhoid have been eliminated, and health care is provided by the county doctor. Young people are required to attend school through nine grades, and most continue through high school to try to pass entrance examinations for the universities. A local cooperative, which provides credit at low interest rates, helps the farmers procure fertilizers and seeds at reduced prices and sell some of their produce directly, eliminating the high profits for the merchants.

Draft animals have been replaced by combines, farm trucks, and tractors with labor-saving attachments to cultivate row crops. Improved breeds bred by artificial insemination far outproduce earlier primitive stock. Irrigation and flood control projects in the villages near the river have reduced damage from spring flooding and brought water to the fields during the hot, dry summer months. New crop varieties and hybrids along with new cultivation practices have increased productivity and income for the farmer. Extension agents, home economists, veterinarians, and Agricultural Bank technicians visit the village to give advice. Village leaders take part in periodic planning meetings with the local prefect or one of his representatives and meet regularly to discuss their own problems at the village level, which has given them all a stronger sense of self-confidence.

Similar progress in varying degrees can be observed in most villages throughout Greece, especially in the plains areas that have exploited their potential for increasing agricultural production by introducing new crops such as early vegetables for export, oranges, peaches, apples, sugar beets, or cotton. The population in mountain villages that depended on livestock grazing, tobacco cultivation, and other labor-intensive crops has decreased considerably, although tourism has played an important role in revitalizing many depopulated island and coastal areas. Among the villagers who have remained in isolated regions incomes have increased significantly, especially during the 1980s as a result of European Common Market subsidies.

The layout of the villages has followed the traditional pattern. The central square with the public buildings including the church, community offices, the school, and a few stores and coffee shops form a hub surrounded by the village homes, storage facilities, and barns. On the outskirts of many villages small factories, processing plants, and rural industries, often established by returning emigrants, provide supplementary occupations for the residents. The fields, which are broken up into small plots that seldom exceed two or three acres, extend in every direction beyond the village. One of the major problems in Greece today is that the average farm is composed of less than eight acres and that it is fragmented into as many as seven or eight plots. Frequent bus connections, telephone services, and private vehicles have revolutionized communication among these villages.

THE IMPACT OF THE PEASANTS

What brought about such a dramatic change in such a brief period? Professor William Hardy McNeill of the University of Chicago, who lived in Greece immediately following World War II and studied changes in six Greek villages in consecutive decades from 1946 to 1976, gives an excellent analysis of the various factors accelerating development.[6] This book, however, deals primarily with the contribution that management training can make in changing peasant attitudes and helping them develop self-confidence and does not allow for a detailed analysis of the economic and historical factors that accelerated development.

The peasants who have moved to the city since World War II speak about the joys of village life, but few would have stayed under conditions prevailing when they left. It is difficult to condemn them for wanting to leave when they did. Development workers in Third World countries should identify the changes in Greece during the past quarter century that have made life in the villages sufficiently attractive for peasants to prefer to stay. Clearly factors such as better roads, improved communications, more effective marketing systems, rural electrification, and rural industries[7] have played a significant role as have the congestion and stress of city life. Rather than migrate to the pigeon-house apartments of the industrial centers many farm families prefer to remain in the rural areas and are optimistic about their future there.

The background of the Greek peasant, as well as his proximity to more advanced countries, made him uniquely receptive to rapid development in a number of ways. First, a thread of continuity was kept alive by an evolving culture, strong ties to Greek Orthodoxy, and distinctive elements in the Greek language that made him conscious of his heritage—what he refers to as "the glory that was Greece." McNeill emphasizes a second factor, the market economy in the villages during the four hundred years of Turkish occupation, that helped provide a training ground for managers in succeeding generations.[8] Some villagers also acquired an adventurous spirit prompted by dire poverty that led them to leave their mountain villages to seek their fortunes in nearby districts and faraway lands. The more enterprising young people from island and coastal villages often risked their lives on unknown seas, hoping for better prospects. But the fact remains that at the turn of the century and even at the beginning of World War II more than two-thirds of the population lived in the villages that had changed little over several generations.

Training too has played a role in rural development throughout the postwar period. A corps of progressive farmers were given useful advice by extension agents and home economists or attended short-course centers located throughout Greece. Graduates of vocational agricultural schools profited from institutional instruction that accelerated the learning process and inspired confidence and a feeling of optimism. The American Farm School has been one of a number of institutions training master farmers from various parts of the country. Although it is virtually impossible to transfer programs from one country to another, many of the activities outlined in this book should prove helpful to others attempting to organize agricultural development programs in countries of the Third World.

3

The Farm School Model

- Are there dimensions to agricultural schools other than training programs?
- Is there a place for adult education in secondary agricultural schools?
- What are the most useful methods in management training?

One day a beggar smelled the appetizing odors of a goat being cooked on the spit outside a rich miser's house. He sat down on a nearby stone and started eating his crust of bread, dreaming that he too was eating a roast, as he sniffed the goat. Just as the beggar was finishing, the rich man spied him and demanded payment for the smell of his goat. Following a lengthy argument the two were brought before Hodja, the judge. When he heard the story Hodja asked the beggar if he had any money. Protesting, the poor man finally pulled out two coins to hand over. At that moment Hodja withdrew his hand and allowed the coins to drop on the floor. He then asked the rich man if he had heard the coins dropping on the stone floor. When the miser said he had, Hodja paused for a moment and replied, "May the sound of the coins be payment for the smell of the goat."

Dr. John Henry House, a resolute American missionary, founded the Farm School in 1904 after working for thirty years in the Balkans. The more he worked the more he realized that what the people really needed was to learn to earn their daily bread. He was a practical idealist who saw the need to give young people from the villages more than just knowledge; he wanted to train the whole individual to bring about what is referred to today as a metamorphosis in their heads, their hands, and their hearts and to develop a sense of commitment and service to their fellow beings. He recognized the importance of the "sound of the coin" to a peasant but was equally anxious to emphasize the quality of village life reflected by the "smell of the goat." Dr. House had no money, but he had a rock solid faith. One visitor referred to him "as a man on fire." With borrowed funds he and his wife built a two-room mud-brick house with space for twelve orphan boys whom he brought up to understand that there was dignity in working with their hands. Dr. House taught the students tirelessly and set a pattern of introducing new methods and new equipment, a practice that continues today. As he had no money to pay a staff, he taught English to a carpenter, a mason, and a cobbler; in return they taught their trades to the students. From the very beginning, however, he regarded the school as essentially Greek and called it the Thessalonica Agricultural and Industrial Institute. The title American Farm School was used by the Greek people to distinguish it from the nearby University of Thessaloniki farm.

In 1916, when Dr. House was seventy-one years old, the first classroom building was completed, but within a year it burned to the ground. Dr. House's son Charles came from the United States for a year to help him rebuild and eventually succeeded his father in 1929 as director, staying at the school for thirty-seven years. As an engineer he was able not only to assist his father in administration but also to supervise the construction of a modern campus as well as a model farm with extensive livestock and poultry facilities.

The original cottage on fifty acres of barren land has been developed into what can be referred to as an oasis—three hundred and seventy-five acres with fifty buildings and a staff of almost one hundred. The original twelve students are now two hundred and twenty boys and girls in a three-year high school level course, more than three thousand men and women attending short courses sponsored by the Greek Ministry of Agriculture, over ten thousand visitors each year from within Greece as well as from abroad and over twenty-five hundred graduates spread out in the villages throughout Greece. Located on the outskirts of the city of Thessaloniki, Greece's second largest city with a population of approximately eight hundred thousand, the school is laid out to resemble a modern Greek village with residences, school, church, and shops clustered in the center surrounded by fields and pastures.

As discussed in Chapter 1 an agricultural school's objective is to train the student to become a master farmer able to manage the new agricultural technology. The Farm School history provides a rich store of experiences that are used in this chapter to illustrate the principles involved in this training: balancing theoretical and practical training, developing a management training program, and utilizing short courses to work with adults.

It is interesting to note, from a list prepared at the time the school was founded, what Dr. House considered necessary to start his school:[1]

1. A house and stable.
2. A brick oven for baking.
3. A horse and two-wheeled cart for hauling provisions, etc., from the city.
4. One yoke of oxen for the winter plowing and another yoke for the spring.
5. Two cows for milk.
6. Plows and other farm implements.
7. Kitchen utensils, furnishings, bedding, etc.
8. Wages of farmer in charge.

In addition to classroom study, the Farm School used an apprentice approach to develop the students' practical skills. Dr. House also realized the importance of developing proper attitudes among the students. One graduate tells the story of having built a wall that was not completely straight, though as far as the student was concerned it was straight enough. When Dr. House saw it he made the boy tear it down and start over. He has never built a crooked wall since.

The House family took great care to kindle self-esteem in the students. They developed a personal interest in each student and took pleasure in discussing the students' concerns and ambitions with them. Dr. House felt that agricultural training, dealing with problems close to nature and to God, would train a boy for the vocation of his choice, whether it was medicine, law, or agriculture.

Dr. House chose as the school's motto, *Laborare est orare*—To work is to pray. It is clear from discussions with graduates of those early years that the students were inspired by Dr. House's ideals of service to one's fellow men, and they acquired his personal love for nature and farming. A young visitor in 1920 observed a certain aura about the school:

> As I came into town yesterday morning I realized that I never before had been in such a literally heavenly place. I'm not using slang. The place, the people, and the work they are doing combine to make one feel good all over and thru and thru. You can't tell just what does it, but the spirit of the institution grows upon you and fills you with a great peace. Either the place is perfect or I am temporarily blinded.[2]

THE YEARS OF EXPANSION

Charles House had graduated in engineering from Princeton University and was especially interested in the students' technical training. When he succeeded his father as director in 1929, he reduced the training program from five to four years and changed the language of instruction from English to Greek. Only applicants who owned land were accepted. Half the students spent each morning in the classrooms, whereas the others were divided equally between the agricultural and industrial departments. In the afternoon they changed places. He felt, as his father had before him, that the boys would learn from example. One of the school's favorite stories is about the official who came looking for Charles House and was directed to the shops. When he arrived he saw a pair of legs sticking out from under a car and said, "Can you tell me where the Director is?" The person under the car replied, "What do you want him for?" The visitor told him that it was none of his business. At this point Charles stuck his head out from under the car and said, "I am the Director. What can I do for you?"

Each department of the school was organized as a separate accounting unit and was required to pay its own way as well as to contribute to the school's income. Charles House believed that it was important for the students to learn not only the practical skills in the departments but the management details. He felt that the only way to learn to manage a profitable enterprise was to work in one.

In his advanced years Dr. House had contended that if he had to do it again he would have trained girls rather than boys. "If you train a boy," he said, "you train an individual. If you train a girl, you train a whole family." In the postwar years Charles House and the British Quakers fulfilled his dream by starting a school for village girls near the Farm School.

In 1946 short courses for adults were started in cooperation with the Greek Ministry of Agriculture. A farm machinery shop and a canning center were established in a unit that housed forty trainees. This represented a new approach to training master farmers as well as village women. These activities were integrated with the Ministry of Agriculture's new home economics program.

One of the primary concerns throughout the school's early years was to find and train staff members who possessed both practical skills and theoretical knowledge. Programs were developed to send abroad young people, mainly graduates of the Farm School, both for short-term training and for full university courses. Because trainees who studied abroad for extended periods suffered cultural shock when they returned, it became school policy to limit the training to two years.

During Charles House's tenure the physical plant of the Farm School was considerably enlarged. A separate dormitory building was constructed in which to house and feed students. Agricultural demonstration units in poultry, dairy, and hog farming and in horticulture were introduced. An industrial quadrangle was added that contained shops for training students in electricity, machine shop work, carpentry, plumbing, and painting. New staff housing and additional athletic areas for the students were completed. From a small school with limited facilities the Farm School was developed into a comprehensive educational institution.

PROGRAM DIRECTIONS

From 1955 until the present, the school has expanded its program in a number of areas. Although it is difficult to separate one aspect of the program from another, each plays a separate role in the development of the master farmer. When Charles House retired, the Board of Trustees undertook a comprehensive survey of the school and the work of its graduates in order to assess the impact of the school on Greek agriculture. The survey included a summary of the graduates' occupations in 1955:

> On the basis of a 10% sample of all graduates since 1927, it appears that 41% of the graduates go into farming in Greece; another 28% have gone into various agricultural services. This makes a total of 69% in farming or related services. The rest are either on scholarship in the United States, have emigrated, are unknown, or are in non-farm occupations.[3]

The most recent graduate survey in the late 1970s indicates that fifty-eight percent of the graduates are employed in agriculture and related services, reflecting changes in the country as a whole. Between 1950 and 1980 the farm population has decreased from sixty to twenty-six percent of the total population. The opportunities for further study open to farm school graduates and the appeal of urban positions have also played significant roles. The proportion of graduates in occupations related to agriculture has increased relative to that for graduates operating their own farms.

The Need for Equivalency

A major complaint of the graduates in 1955 was that their graduation certificate was not equivalent to the diploma issued by the Ministry of Education. For years Charles House had avoided establishing a diploma that would relate to that for Greek classical education because too much time would have to be devoted to academic subjects at the expense of agricultural training. He also feared that if students obtained an equivalency diploma the large majority would go on to higher education and not return to their villages. This lack of an official diploma led to a decreasing desire among the villagers to enroll their children. Obtaining an education, particularly an accredited education, is the greatest ambition of all Greek parents for their children. It was a major disappointment to them to have their sons go through four years at the Farm School without any recognition beyond a grammar school certificate.

The 1955 survey helped to clarify the goals of the school, the needs of the students, and the programs that would meet these needs. It was important that the school should keep up to date with both the changing agricultural scene and the expectations of the students and their parents. Seeking accreditation was one of the principal recommendations of the survey. In 1963 a three-year gymnasium equivalent program was incorporated into the four-year course so that the students would have a degree equivalent to the one awarded in urban institutions. In 1973 the level of the school was raised from junior high school to senior high school, offering courses to train technicians to use and maintain farm machinery, to breed and raise livestock, and to grow and market crops. Finally, in 1978 the school was recognized by the Ministry of Education as a fully equivalent vocational lyceum. This equivalency was a great source of pride to the alumni who had graduated during the old program. They no longer felt inferior to urban high school graduates.

The New Approach

Management training became a focal point in the revised program. Some of the staff who had met members of the Future Farmers of America (FFA) were very impressed by their self-confidence, their public-speaking ability, and the assurance with which they managed their home farms. The staff felt challenged to develop comparable young men and women who held their heads high and were proud to be farmers.

Members of the FFA gain practical experience by managing agricultural enterprises on their own farms to supplement their classroom instruction. This approach is not very different from the Harvard Business School's case-study method. The Farm School's administration decided that if this technique worked in the United States it could be adapted to the needs of village boys and girls in Greece. The school became an agricultural bank and extended credit to groups of students who raised such produce as pigs, chickens, cows, vegetables, and field crops. Students were required to prepare budgets, keep records, and manage their enterprises effectively. With their profits they were able to take field trips at the end of their senior year.

The students' interest in making more money developed into the primary motivating force in teaching them how to be effective managers. The small farm enterprises, known as student projects, became the backbone of the school's management training program. Students were also involved in the production departments, in which they were paid as laborers. By the time they reached their senior year those who had acquired the necessary skills were made responsible for the management of whole departments on weekends. The school's staff is still trying to improve management training. Ideally, students working as members of a small team should have a few projects for which they must assume sole responsibility at school or in the village.

The school has designed a number of one-person units—the optimum enterprise that can be run by a farmer and his family. These units serve as demonstrations for the short courses and training areas for the regular students. In each of these, careful records are kept that are the basis for management instruction.

Running short courses for adults over the past thirty-five years has confirmed that they should be a basic part of any agricultural institute. They benefit not only the trainees but the staff members who come into contact with the trainees. Teachers become involved in the concerns and problems of the farmers, which ensures more stimulating instruction for the regular students and avoids the danger of stagnation. Short-course trainees continually challenge the school to keep its programs up to date and relevant to the actual needs of the villages.

Another focal point of the school's curriculum has been innovative technology. The school has taken the lead in introducing agricultural technology in a number of fields. This does not mean importing the latest U.S. or European equipment but rather adapting the technology developed abroad to Greek conditions (see Appendix A). The school has developed more than seventy such innovations, the most significant of which include the reaper and binder in 1914, the first pull-type combine in 1935, and the first pasteurizer the same year. The incorporation of this technology in training programs has stimulated staff and students to seek new solutions to old problems and innovative approaches to current concerns. The latest focus is on solar and other energy sources, and a small alternative energy laboratory has been set up to involve the students.

People visiting the school become aware of an intangible dimension beyond the structure and function of the various departments. It is referred to by the staff as the "spirit of Dr. House"—the quality of the school. In some institutions this quality can be sensed in the personal involvement and dedication of the teachers for the students, the warmth among the students themselves, and a spirit that at the Farm School expresses the founder's inspiration.

Influences Outside the School

Integrating the program more closely with the Greek government through the Ministries of Agriculture and Education has been one of the major objectives of the school in recent years. Experience in European and developing countries indicates that agricultural schools supervised by the Ministry of Education tend to use a theoretical approach and to be removed from the practical problems of the farmers. On the other hand, agricultural schools under the Ministry of Agriculture lack the much-prized formal equivalency. For the Farm School this difficulty has been overcome by establishing a joint Ministry of Agriculture and Ministry of Education committee, which is responsible for supervising the training program. The Ministry of Agriculture also provides substantial support for the students' training, which leads it to become more involved. A new short-course center completed at the school was financed by funds from the World Bank and the Ministry of Agriculture.

It has become increasingly obvious that women play a vital role in management in the villages. One of the most controversial steps taken by the school in recent years was to combine the training of boys and of girls. Trustees and faculty doubted the wisdom of housing them in separate sections of the same building. Much to everyone's surprise none of the parents seemed to have any objection, and the students themselves thought it quite natural.

The school has been less successful in organizing joint courses for men and women in the continuing education program. Agriculturists feel more comfortable training men, and home economists prefer to work with women. A future challenge will be to integrate this instruction.

The increasing number of trainees from Third World countries who come to study development in Greece has prompted the school to seek ways of sharing the Farm School experience with others. The fact that Greece has progressed so remarkably since World War II makes it a far more suitable training ground than the more advanced Western countries. Although the total concept of the Farm School cannot be transferred, many principles and specific practices could effectively be applied elsewhere.

WEAKNESSES AND SHORTCOMINGS

Many of the Farm School's accomplishments during the last eighty years have been outlined, but the school has also experienced shortcomings and failures. As a private institution, the school has difficulty attracting permanent staff because it must compete with the greater financial security that the Greek government offers to agriculturists and others. At the same time, however, it is fortunate to have had a highly qualified and dedicated nucleus of staff members that, despite personnel turnover in a number of areas, enables the school to continue to operate effectively. It is vital for any such institution to train staff constantly, recognizing that though many members may stay for only a few years they often make significant contributions in government service later—applying what they learned at the Farm School in their new jobs.

Another sphere in which more could have been accomplished is in developing initiative in the student projects. Too many students work on each project, and they do not learn proper planning and record keeping. They are eager to earn money but not sufficiently motivated to follow the proper management techniques. Because the concept of projects is still new to Greece, it takes a long time for staff members to learn to supervise them effectively. The turnover in the staff members responsible for student projects disrupts their continuity.

Too much time is given to lecturing and not enough to allowing the students to take the initiative in discussions and in practical work. Despite constant pressure on the instructors to speak less and to encourage the students to work more, they still tend to revert to the traditional Greek classroom pattern.

The school should have more long-range planning. Although a ten-year plan has been prepared, it requires constant reassessment, particularly as changes in the national government policies affect education and agriculture throughout the country. Both as a private and as a foreign institution in Greece the school has always been slightly suspect among those who have not had close personal contact with it. It is vital that the planning of the school be closely identified with the program of the government in office.

Consistency in management has often been a weakness. Although clearly defined objectives have been required from each department, some staff members have been delinquent in preparing them or ambiguous in describing them. Some departments have also failed to follow through on the objectives, particularly in relation to time schedules. In this area staff turnover has again been a major factor. Following up on the implementation of objectives is a constant challenge in institutions such as the Farm School.

Although continuing education has played a very useful role in relating the school to village problems, the courses for adults have often failed to meet trainee needs. In courses in which teaching staff and administrators rather than the villagers dictate content, attendance has been poor and interest only perfunctory. By interviewing and having discussions with the villagers during the planning process, the staff can overcome the tendency to ignore the villagers and include their input in teaching packages. The lack of recognition of the peasants' understanding of village life and needs is a common problem in training centers throughout the world.

Even though the school has made progress in some areas of technology, it is behind in others. So far it has made little use of the computer in agriculture, although a number of microcomputers have been installed in the school library for the students. Because so many innovations have been made in so many fields, it is difficult to keep pace with the changing technology. Tractors become outdated, sprinkler irrigation is replaced by deep irrigation, and computerized incubators replace manually operated units. A piece of expensive equipment that seems advanced one year tends to be obsolete three or four years later. Agricultural companies like Sperry New Holland are leasing equipment such as a new combine to the school at minimal cost, which reduces the amount of capital required.

For many years the school operated an effective graduate follow-up program in which a full-time staff member worked with the graduates in their villages. He helped some to find employment and others to obtain loans from the Agricultural Bank with which to introduce new innovative technology. This position is probably the most difficult to fill, for it requires someone with broad technical skills willing to spend many hours traveling to the villages. The school has only recently been able to find such a person. One solution is to have a generalist maintain regular contact with the graduates and to send appropriate staff members to the villages to deal with specific problems.

Few people, even in the surrounding community of Thessaloniki or in the rest of Greece, have any detailed understanding of the school's program or its objectives and accomplishments. Although most people, when asked, will say it is a good school, they are more inclined to be speaking of the high quality milk, eggs, and turkeys that it sells than of its educational program. The school requires a more effective public relations program to make known the variety of educational and technical programs that it operates.

Finally, the hands-on training of the students deserves further study. Although members of the teaching staff have prepared a list of the basic skills that the students need to learn, they have not yet defined the specific activities through which the students develop these skills. The school was very fortunate to obtain the assistance of an expert in curriculum development and vocational education to guide the staff. However, one or more faculty members should be specifically engaged in preparing teaching packages and audiovisual aids, as well as translating available material from other languages. It is unrealistic to expect a busy faculty in a many-faceted institution like the Farm School to have time to prepare such specialized material.

In this chapter a summary of both the positive aspects and the weaknesses of the Farm School programs has been presented. It indicates how imperative periodic evaluation and review are to an institution. Organizations, like human beings, suffer from a gap between their objectives and the discipline needed to make realities of their dreams. The school has gained much from its own experiences and the success of others.

4

Dynamic Training Centers

- What leads to excellence in an institution?
- What attributes should schools try to develop?
- Can the staff of institutions be both compassionate and businesslike?

One of Hodja's neighbors complained bitterly to him about having too many children and a very small house. Hodja promised to help him solve his problem if he would agree to follow his instructions religiously for a week. Utterly despondent about his situation, the neighbor accepted the terms. Each day during the week that followed the wretched neighbor was told to move a different animal into his already cramped house—the first day his cow, followed by a goat, a donkey, a pig, two sheep, and the mule—until he felt he could stand it no longer. On the last day Hodja told him to remove all the animals and come to see him the following morning. "Glory be to Allah," said the neighbor to Hodja when he returned, "what a delightful house I have."

It is so easy to find fault with almost any training institution. Staff members, trainees, and visitors tend to complain about the facilities, the program, the instructors, the food, or some other actual or imaginary shortcoming. As with Hodja's neighbor, they find the family too big or the house too small. A study by the author in more than twenty countries, however, revealed that outstanding institutions share certain intangible qualities that have a lasting impact on the trainees and visitors, as well as on the staff itself. Invariably, discussions with the staff of these organizations center on the fundamental question of why certain institutions and development programs are conspicuously successful.

These helpful comments and suggestions pinpoint four common attributes of exceptional schools: a clear direction, standards of excellence, a sense of community, and an atmosphere of compassion. These characteristics are particularly relevant to training centers but are equally valid for village-level community development programs. In the following sections, institutions that illustrate these characteristics are described.

CLEAR DIRECTION

The home of one who does not praise it will cave in and crush him.

—Greek peasant saying

Top administrators of any training center must be convinced that its work is vital and that its mission has a sense of greatness about it; only because of these convictions, can the administrators pass their enthusiasm on to others. These convictions permeate the thinking of the administrative staff of the Instituto Superior de Agricultura (ISA) at Santiago in the Dominican Republic. ISA is an agricultural college operated in cooperation with the Catholic University in Santiago. It is directed by Dr. Norberto A. Quezada and a group of young associates—an inspired team with single-minded fervor. Demonstrating the best qualities of dynamic salesmen to fellow staff, students, and supporters, they convey their firm belief in the substantial contribution that ISA is making to their country. Their unwavering ambition for its programs and their impatience with mediocrity are fully shared by the faculty. An extended group of loyal supporters shares this belief in ISA and is inspired by a staff that has attempted to make their dreams of greatness a reality.

Another factor that gives a clear sense of direction to an institution is a well-defined statement of the mission. Lack of agreement among the members of the teaching staff and administration on the goals of an institution often leads to many hours of fruitless discussion, which might more usefully be devoted to program development. Kenneth

Blanchard and Spencer Johnson use the term "one minute goals" to emphasize the importance of precise objectives.[1] They are convinced that goals should be expressed in fewer than two-hundred and fifty words, so that they can be read and understood by anyone in less than a minute. The need for a concise mission statement was discernible in a short-course center attached to an agricultural school in East Africa. There the staff members disagreed so much about the mission of the newly built center that their discussions left them little time to implement programs. Once they decided on a mission, they were able to focus on creative training activities.

An organization with a clear definition of mission is CIAT (Centro Internacional de Agricultura Tropical), outside Cali, Colombia. This autonomous, nonprofit institution is dedicated to international agricultural research and training: one of a group of similar centers located around the world that have contributed so greatly to the Green Revolution.

> The mission of CIAT is to generate and deliver, in collaboration with national institutions, improved technology which will contribute to increased production, productivity and quality of specific basic food commodities in the tropics—principally in Latin America and Caribbean countries—thereby enabling producers and consumers, especially those with limited resources, to increase their purchasing power and improve their nutrition.[2]

CIAT is seeking to develop new varieties and improve production practices of four crops basic to Latin America—beans, cassava, rice, and pastures—by assembling information, making laboratory tests, and operating experimental plots.

If the objectives and deadlines for each program are not specified in a mission statement, the direction of an institution can be ambiguous. CIAT is an outstanding example of an organization that has defined its objectives clearly and that has established specific time frames that are constantly reviewed by the staff and the board. Programs that have failed to meet approved schedules are subject to reappraisal and may be abandoned. This makes the staff members feel a constant sense of urgency often lacking in other organizations.

Homer Lackey, a former steel mill manager, insisted on developing similar well-defined objectives when he was engaged to help the Farm School. At that time there was confusion about the direction of the school and many departments were being run inefficiently. He treated each production section as if it were a steel mill, analyzing its goals and productivity in comparable terms. He then helped the staff reorganize the education departments. By the time he left, each department of the school had clear objectives and agreed about how and when they could be accomplished.

A deep commitment to an underlying ideal—a force behind the program's existence more fundamental than the institution's objectives or structure—also contributes to a clear sense of direction. Such a commitment had a dynamic impact on the Indian Meridian Vocational Technical School in Oklahoma. The assistant superintendent, Leroy Bailiff, had a deep sense of dedication and knew exactly where the school was going and why. The administration had a detailed time schedule for the coming years, which was worked out with the faculty to ensure their full cooperation in achieving the school's goals.

Students of all ages were enrolled in this impressive high school complex. The faculty observed that the older students, who knew what their own goals were could serve as a major motivating force for the younger ones. Students took their basic course work in local high schools and then attended the technical school for three hours each day. The school offered two hundred and eighty-seven evening courses for which instructors were hired from industry. In accordance with the underlying philosophy of the school, the courses were organized to tie in with skills demanded by local industry. Each department had an advisory committee that helped define its goals. Visitors were impressed by the school's strong sense of commitment to outstanding education, apparent in the order and organization, in the involvement of the teachers and administrators, and in everyone's conviction that it was a very special school.

STANDARDS OF EXCELLENCE

According to a Middle Eastern proverb, "The fish starts smelling from the head and the rosebush from the highest rosebud." Although everyone working in a training program should aim for high levels of excellence, the inspiration for a school must begin with the top administrator. If he does not demand excellence of himself, his department heads, and his teachers, then they in turn will not require it of their employees or of the students. Training centers are the home and training ground for farmers and young people. If they are exposed to an atmosphere of excellence at school, they will tend to take it with them when they leave. The staff of an institution must tangibly express the values that it is attempting to inspire in the trainees rather than conceiving of the school as a place that is preparing them for a future occupation unrelated to their present environment. The sentence "Education is living, not preparation for living" is engraved at the entrance to the library of the University of Rochester.

These same high standards prevail at ISA in the Dominican Republic. Besides being concerned about the personal welfare of the students, the director and his staff are also strict with them, in both their academic work and their practical instruction. Only when it becomes very clear that students are not responding to these challenges are they encouraged to transfer to another school. One of the difficulties in vocational education is to set high standards for the capable students while also being supportive of those less motivated.

Excellence in an institution has little value unless it percolates down to the students themselves. The staff must ensure high levels of competency among the trainees through their own activity in accordance with the principle, "I do, I understand." At the Panamerican Agricultural School in Zamorano in Honduras (Escuela Agricola Panamericana) a sense of the work ethic is unmistakable. The students and staff know that in the practical departments they are expected to work, not just to learn.

In such an atmosphere students have difficulty distinguishing between learning and gainful employment. Working to learn is a self-centered process in which all a student's activities lead to acquiring either knowledge or skills. In contrast, when employed the student gives something of himself in return for financial gain. The Panamerican Agricultural School instills in the students a work ethic that effectively combines learning with productive work. Graduates are easily identified in Latin America by their dedication and their ability to deal with practical problems—attributes that have grown out of the habits they acquired during their training.

Throughout Switzerland the agricultural schools share the same work ethic. The students obviously take pride in the high quality of their achievement, manifested in the cleanliness and order of their shops, in the way their farm operations are organized, and the thoroughness of their farm management studies.

Teaching excellence demands a flexible, innovative approach. Over fifteen years the Farm School has organized four new programs, each of which has proved to be increasingly effective. One of the most exciting aspects of the agricultural school at Emmeloord, Holland, is its regenerative spirit. Both the farmers and the instructors tend to be unusually receptive to change and innovation. This openmindedness results in part because their school is new, built on recently drained lands.

It is vital that institutions involved in development work retain staff members who think in terms of the "why" of their programs and not just the "what" and the "how." The great U.S. land-grant colleges and universities have kept abreast of the needs of farmers because of the innovative spirit of many of their staff. Key faculty members at such schools as the University of California at Davis, Cornell, and Michigan State University are specifically concerned with creativity, problem solving, and innovation in dealing with farmers' problems.

A sense of order and orderliness can definitely contribute to the feeling of excellence in an institution. The Indian Meridian Vocational School had an almost intimidating sense of order. Every part of the school was color coded, so that a visitor only needed to follow a particular color to reach a destination. Each shop, laboratory, and classroom had its own rationally determined floor plan, and each laboratory submitted a daily report on lost tools. It was impressive that a school with such enormous resources was at the same time so meticulous about the loss of even the smallest tool. Maintaining control of facilities is a difficult aspect of managing training centers in developing countries because order and orderliness are not always valued very highly in a particular culture.

Attention to orderliness is very apparent at two schools operated by the Silesian Fathers in the Dominican Republic and in Ireland and at An Grianan, near Dublin, which is operated by the Irish Women's Association. Cropped lawns, carefully kept gardens, neatly painted doors and windows, spotless walkways free of paper scraps and cigarette or candy wrappers, and attractive entrance ways to all three institutions created a favorable impression.

Long-range planning and budgeting also play a role in creating a sense of excellence. CIAT and the Indian Meridian School each have a five-year plan. Larger institutions like Michigan State University and the University of California at Davis use both five- and ten-year development plans. Long-range planning is equally essential for relatively small institutions such as ISA and the Farm School, which maintain such plans both for their school and for the outreach programs in the surrounding areas.

SENSE OF COMMUNITY

A sense of community among staff and trainees and the recognition of this by outsiders are at the heart of a successful rural development program. A definition of community, attributed to Peter Pongis, former general secretary of the National Foundation of Greece, describes this sense as "the area circumscribed by the sound of the village bell, calling the faithful to worship, to a baptism, a wedding or a funeral, where the happiness of one is the happiness of all, the sadness of one is the sadness of all, and the secret of one is the secret of all." This feeling of shared happiness and concern between staff and trainees builds a lasting bond between them and reinforces the educational programs.

The campus of the Farm School outside Thessaloniki has the appearance of a village, and both staff and students feel themselves to be a part. The visitor can hardly fail to notice this at An Grianan (which means "on the sunny side of the house"). Ann Powers, the director, suggested some essential requirements for effective management that reinforce this feeling:

1. Comfortable classrooms and facilities.
2. Good food served in a friendly manner.
3. Hospitable and clean sleeping quarters (single, if possible).
4. Familiar symbols, such as county names, for identification.
5. An introductory session so that everyone feels welcome.
6. A hostess for each course to act as housemother to the group.
7. Name tags for identification.
8. Activities for every evening of the week, such as Monday, introductions; Tuesday, lecture; Wednesday, free night; Thursday, recreation; and so on.
9. A center that operates efficiently without losing its friendly atmosphere.
10. A staff that makes the trainees feel at home.[3]

A special dimension of warmth—expressed by the Silesian Fathers in the Dominican Republic as *bondad*, by the Greeks as *kalosini*, and the Tanzanians as *ujamaa*—adds immeasurably to the feeling of community. The more the trainees sense this the more they are able to relate to the center, which motivates them to return for further training. Warmth and affection pervade the whole campus at the Mtwara Agricultural School in Tanzania, a training center in one of the poorest agricultural areas in the world.

A positive approach to discipline by all levels of the staff, which emphasizes merit rather than weakness, strengthens a sense of community in a school. Ann Powers commented that a kind word is as easy to say as a nasty one. However, there is an unfortunate tendency in training centers for some staff members to be harsh and patronizing to farmers and farm youth.

The opposite approach is demonstrated at ISA by the dean of students and the head of the dormitory. They regard themselves as an older brother and sister to the students rather than as superiors or disciplinarians, and they set an excellent example for the rest of the staff. This approach is also used by the director of the Warrentown Agricultural

School in Ireland, who thinks of himself more as a parent to his boys than as a teacher.

A problem-oriented rather than person-oriented administration is an important factor in avoiding conflict within the educational community. Staff members should learn to concentrate on problems that arise and to play down personal animosities. Once an individual's feelings are allowed to intrude, problem solving is seriously disrupted. Over the years, many Farm School problems have arisen from antagonism among staff members. These have wasted a lot of time and often interfered with the training program. A development worker in Tanzania described a staff meeting in which three hours were spent resolving a problem that could have been solved in about fifteen minutes. The delay resulted from the staff members working at cross purposes rather than from the complexity of the problem. When in exasperation he pointed out to staff members that they were trying to undermine one another, they were able to focus on the problem and solve it.

It is virtually impossible for a staff of clock-watchers to create a community feeling in training centers that necessarily operate twenty-four hours a day. At the Farm School about twenty-five percent of the staff members are so deeply involved that they are prepared to give their time and energy unstintingly to the school. Another fifty percent work hard during their eight-hour day but are not particularly anxious to work overtime. The remaining twenty-five percent only give as much as is required of them. These percentages are probably representative of staff commitments in most training centers.

However, there appears to be a complete sense of twenty-four-hour involvement by the Silesian Fathers at their centers. They act as fathers, mothers, brothers, and sisters to the students. The students know this, and the institution reflects it. At the Panamerican Agricultural School in Honduras the director and staff live on campus and are therefore always available to the students.

ATMOSPHERE OF COMPASSION

A faculty with a sense of compassion, which shares the concerns and problems of the trainees, is an indispensable aspect of a good training center. At the Farm School the students feel that they are an integral part of a community that is truly interested in them and concerned about their welfare and personal problems. When students seek help they must have the full attention of the staff. "Shared happiness, twice the happiness; shared pain, half the pain" is a common Greek saying.

At ISA in the Dominican Republic a young and dynamic staff, many of whom have masters degrees from abroad, express this sense of compassion in their concern for the students. They know each by name and understand their problems, and this concern brings out the best in each boy and girl. This sense can also be found in the short-course center at Emmeloord, where courses are run for groups small enough to allow the teacher and the trainee to interact and relate to each other. An outgrowth of this feeling of compassion is reflected in a lightness of touch and humor, in contrast to the institutional tone, which tends to be heavy, rigid, and conservative. Instructors often lose sight of their responsibility to serve the trainees rather than vice versa.

Theo Litsas, the Greek associate director of the Farm School, was for many years the heart of the institution. His greatest asset was that he was totally involved with every person associated with the school—staff members, trainees, and visitors. When he was killed in an automobile accident, a Greek trustee pointed out that Litsas had brought a lightness of touch and a sense of humor to the campus—both of which added an extra dimension to outstanding development programs.

Compassion is also manifested in an organization's concern for the whole individual: the mind, body, and the spirit. Anyone who has visited a good Young Men's Christian Association (YMCA) program anywhere in the world has observed how important it is to cater to all aspects of the human being. A staff member at the Farm School resigned after three years: When asked why he was leaving he replied that the Farm School was too much like a perpetual YMCA summer camp program. This is perhaps one of the greatest compliments that the school has ever received.

Just as humility is a characteristic of many great people, so the measure of a successful institution is a staff with a high intellectual capacity combined with a personal concern for the trainees. In addition to the academic excellence of the faculty of schools like the University of California at Davis, Michigan State, and Oklahoma State, the members are sincerely interested in helping the casual visitor and learning from him. No doubt such professors have a lasting influence on their students as well as on others. Unfortunately there is a dearth of this caliber of dedicated staff members in developing countries.

Training rural people to become capable managers is not an easy task: It involves more than just teaching them the elements of leadership and related skills. Dynamic institutions require leaders that give them a clear direction and demand high standards of excellence. At the same time they must create a sense of community that leaves the trainee no doubt about the deep individual concern of the staff members. Only through personal example can the administration and staff combine these qualities in the management of institutions seeking to train master farmers.

PART TWO

Training the Managers

The most significant common characteristic of the nikokiris, the master farmers, and the sergeants of agriculture is their competence as managers. The key task of development programs must therefore be to help peasants understand the concept and acquire the skills to become competent managers. Because no word equivalent to *management* exists in Greek or most Third World languages, the Farm School staff introduced the acronym POLKA to represent the five key aspects of the managing process—planning, organizing, leading, controlling and adjusting.

When a group of dairy farmers attended a special short course in dairy management at the Farm School, they were asked if they knew how their cows related to the POLKA. They knew that a polka was a Polish dance, but they did not see how it related to their cows. Once they had the acronym explained to them and they learned in the course the importance of the five management elements in their dairy operations, they began to appreciate the concept. A year later when the same group of trainees were asked at a refresher course what they still remembered they replied, "POLKA, POLKA."

The five aspects of the POLKA are essential to the management of training centers. Based on experiences at the Farm School and other agricultural training institutes, Chapters 5 to 9 are devoted to showing how the basic elements of the management POLKA can be implemented both in managing institutions and in training peasants to be better managers.

Hodja was determined to be decisive and efficient. One day he told his wife he would plow his largest field on the far side of the river and be back for a big dinner. She urged him to say, "If Allah is willing." He told her whether Allah was willing or not, that was his plan. The frightened wife looked up to Allah and asked forgiveness. Hodja loaded his wooden plow, hitched up the oxen to the wagon, climbed on his donkey, and set off. But within the short span of a day the river flooded from a cloudburst and washed his donkey downstream, and one of the oxen broke a leg in the mud, leaving Hodja to hitch himself in its place to plow the field. Having finished only half the field, at sunset he set out for home exhausted and soaking wet. The river was still high so he had to wait until long past dark to cross over. After midnight a very wet but much wiser Hodja knocked at his door. "Who is there?" asked his wife. "I think it is me, Hodja," he replied, "if Allah is willing."

5

The Planning Process

- Why is planning important in development?
- Can peasants really learn to plan?
- Is "leaving it to the gods" the most effective approach when there are so many unknowns?

Planning is the process of deciding in the present what will be accomplished in the future. Some managers insist that plans will be implemented whether "Allah is willing or not." Wise managers include two additional steps in the planning process—controlling and adjusting (discussed in Chapters 8 and 9)—to provide flexibility. Peasants learned long ago that even the best plans depend on factors outside their control, such as the vagaries of nature and market forces, and they are inclined to cite their own experiences as excuses to avoid having to plan. For the same reason planning remains one of the weakest sectors in organizations involved in rural development. Staff members, like many of their peasant friends, notice that their institutions continue to make progress despite inadequate planning and therefore wonder why they should spend much time on this aspect.

LONG-TERM PLANNING

Planning is an ongoing process rather than an attainable objective and requires continual review and evaluation. This chapter uses a stormy period in the Farm School's history to illustrate the importance of planning to an institution and the various steps involved. Plans grow out of a clear understanding of an institution's mission, which gives it direction. However, the plans must be specific and clarify priorities and deadlines. In addition to long-range planning, institutions, individual departments, and staff members must have clearly defined medium- and short-term goals through which to evaluate their progress.

The second part of this chapter describes the achievements of one hundred and sixty-five communities in the Thessaloniki Prefecture in planning specific projects and activities for their villages as a part of the Thessaloniki Community Development Program. Another example of fruitful planning was a poultry program to introduce broiler raising among twenty-five Farm School graduates. Peasants need help in clarifying and defining goals and relating them to their own needs—a process described in the final section.

The Need for Objectives

The Farm School's recovery from a period of considerable difficulty in its history demonstrates the importance of careful planning. The school's problems became acute in the 1970s when the Greek government completely revised educational programs and required the Farm School to follow suit. During this period enrollment dropped by two-thirds, and the cost per student more than doubled. Staff members became demoralized, trustees grew disillusioned, and students wondered about their future.

A major weakness in the school's policy was the failure of the director, the faculty, and trustees to clarify their objectives. They wanted to train village youth and adults as well as operate a demonstration farm, but beyond these points they showed little agreement on the school's mission and its long-term goals or on their implementation. Like Hodja, they could have said, "We think it is us, the American Farm School, if God is willing."

As the board's concern about the future of the school grew, various trustees sought different approaches to effective planning, insisting that the faculty clarify their objectives. Marathon meetings lasting into the night were organized at the school. A village friend, representing a peasant point of view, commented that if the faculty had not spent so much time on planning, the program would already be organized.

After considerable discussion the staff agreed that some of the advantages to planning are that

1. Staff and trustees agree about the school's directions.
2. Each staff member understands what is expected of him and his colleagues.
3. Trainees know what to expect of the institution and what will be demanded of them.
4. Everyone associated with the school has a clear picture of what the school is trying to accomplish and what support is required.
5. The existence of objectives makes it possible to periodically evaluate accomplishments.

John Henry House clearly stated his goals in the school's Charter of Incorporation: "providing agricultural and industrial training under Christian supervision for youth . . . in order that they may be trained to appreciate the dignity of manual labor and be helped to lives of self-respect, thrift and industry." Seventy years later, the school's activities and programs had grown too numerous and complex to be defined by such a simple statement. As efforts to resolve the conflict continued, three procedures to provide more effective planning were agreed on: a clear mission statement, a survey by an outside expert, and a long-term view by management.

During the discussions among board members, staff, and others about the future direction of the Farm School in 1978, the chairman and a faculty committee drafted a statement that attempted to summarize where they felt the school should be going. This statement in turn was reviewed by other committees, modified, and finally brought before the

THE FARM SCHOOL'S MISSION

The primary mission of the American Farm School is to provide vocational training for young Greek men and women as well as for adult farmers and their families on a highly specialized basis. In essence, the purpose of that training is to equip them to manage agricultural enterprises or otherwise assume positions of leadership in Greek agriculture.

Among the major programs are educational sequences for boys and girls of high-school age. Each of these programs conforms to the new Greek lyceum/technical school system which the School itself influences through direct training methods and by example. The School also offers a wide variety of extension courses for adults with primary emphasis on advanced farming and craft methods, rural development and management techniques consistent with the mission.

The many other ways in which the School carries out its mission include individual lectures, demonstrations, seminars and, in the future, publications. In its efforts to meet the needs of farmers in various regions, the School plans eventually to deliver programs in communities throughout the country as well as on the campus of Thessaloniki.

As in the past, the School is committed to playing a constructive role in the furtherance of understanding and friendship between the Greek and American people. The School intends to maintain its traditional identity as a non-profit educational institution, with support from the private sector ensuring its independence.

whole board during a two-day meeting devoted to long-range planning and goal clarification. This Mission Statement was adopted after lengthy discussions. Because the staff looked upon this statement as a broad guideline but not as a basis for scrutinizing specific activities and programs, they restructured it in outline form (p. 55) to give precise guidance on programs. The simplified form made it easier to clarify which activities were most important, which the board felt should be emphasized, and what the overall direction of the school should be.

Program-Oriented Planning

While the Mission Statement was being developed, the board commissioned Professor Irwin Sanders, a sociologist and leading authority on rural Greece, to study the objectives of the school in view of the

The Mission of the American Farm School

I. The overall objective is to provide vocational training for rural youth and adults
   - A. To manage agricultural enterprises
   - B. To provide leadership in Greek agriculture

II. Specific Programs
   - A. Middle-level education programs
     1. To train boys and girls
     2. To influence training programs
        - a. Through direct training
        - b. Through example
   - B. Short course training programs for adults
     1. Advanced farming
     2. Craft methods
     3. Rural development
     4. Management techniques
   - C. Other programs
     1. Individual lectures
     2. Demonstrations
     3. Seminars
     4. Publications
     5. Training in other communities

III. Secondary objectives
   - A. Further understanding and friendship between Greek and American people.
   - B. Maintain identity as a private, nonprofit institution with support from the private sector (as well as the public sector), to ensure its independence.

rapid changes in the villages. After interviewing trustees, faculty, officials, alumni, and villagers Dr. Sanders took a different approach in his report. Rather than preparing a broad mission statement that might be confusing, he made his recommendations in terms of specific program goals and activities.

Dr. Sanders then prepared a time frame for staging his recommendations, which provided both priorities and deadlines. The Sanders report concluded:

> Since it is impossible to meet all of the challenges confronting the School at this time, it seems best to deal with them in stages, or in sequence.

Summary of Dr. Sanders's Recommendations

1. Two-year vocational schools in mechanized farming and in rural economics and management, as well as a vocational lyceum in farm machinery.
2. Emphasis throughout on farm management skills, decisionmaking, record keeping and cost analysis.
3. Increased emphasis on teaching English.
4. A Curriculum Materials Development Program.
5. A well-rounded program in continuing education.
6. Production demonstration units to produce revenue and train students and adults.
7. Evaluation of existing production demonstration units.
8. Continuation of existing international programs.
9. Development of a plan for making the school's program more truly international.
10. At a later date development by the staff of a plan for making the School a more truly international center. (A decision on Sanders's Recommendation 10 was postponed until a future date.)

> Staging means that at a given point in time we will seek to upgrade one program while putting other programs on "hold." Administrative and staff time is so limited that to diffuse it too widely is self-defeating.
>
> An agreement upon program staging can make possible long-term budgeting and some predictable allocation of resources over the next three, four or five years.[1]

It was much easier for the board to review each recommendation and approve or disapprove it once they had the specific recommendations and the staging program. The Sanders report has since been an invaluable guide to the staff in planning activities on an annual basis. The Board of Trustees decided to review the whole study every few years to evaluate achievements and to provide further future guidelines.

A third approach to long-term planning implemented by the staff was to look ahead for a ten-year period. This resulted in a view of the institution that was not so concerned with exact implementation dates for specific projects as with broad-brush sketches of programs and the physical plant at the end of the decade. This approach enabled the administration to envisage plant facilities, personnel, and budgetary requirements for the years ahead and submit a ten-year plan for consideration by the board.

Planning Procedures

In addition to the clarification of the school's goals, an invaluable byproduct of these studies was the personal involvement of the trustees and many of the staff members in the planning process. As a result of the meetings and informal discussions they grew closer to each other as well as to the school and its programs. Thus staff and volunteer participation proved to be the most indispensable element of the whole undertaking. What had originally been a near calamity for the school proved to be the focal point of a new spirit of cooperation for faculty, trustees, and other friends.

The most effective way to maintain the interest of trustees is to seek their advice in the planning process, even though this approach may sometimes seem time consuming and frustrating. The same principle applies for organizations that are supervised by government departments or committees of management. It is vitally important to maintain the involvement of those who establish program policy, and it has been said of involved trustees: "More to be desired are they than gold."[2]

Individual staff members feel personally committed to the success of the long-range plan only to the extent that they shared in preparing the recommendations for it. It is better to avoid consulting them entirely than to seek their advice and not allow them to share in the final decisionmaking process. Over the years the richest sources of innovative ideas for the school have been not only faculty members but also laborers and technicians, most of whom grew up in the villages. A large number of graduates employed by the school have played an invaluable role in the planning process. Administrators should never forget the potential contribution of peasant wisdom in developing long-term goals.

Other groups that can make important contributions to the planning process are the students themselves, trainees from the villages, and leaders in agricultural industries. In forming local advisory committees, organizations should select members who represent many associated interests such as the agricultural office, home economics department, district cooperative office, university agricultural education department, agricultural bank, as well as representative dairy, vegetable, and field crop farmers, leading village housewives, and alumni.

MEDIUM-TERM PLANNING

The medium-term planning at the school is presented in a five-year budget. The section for the first year is highly detailed; that for the second outlines the programs that the school expects to implement; the sections for the third, fourth, and fifth years are much more broadly based but still stand up to detailed scrutiny. A flow chart of a five-year plan based on Dr. Sanders's recommendations and a five-year budget based on this plan were submitted to the Board of Trustees. Each year since these were compiled the staff has reviewed the previous year's accomplishments, updated those for the subsequent years, and worked on projections for the final year. The discipline of coordinating program and budget planning has been most useful to everyone concerned and played an invaluable role in the management process.

Probably no management concept has been more generally accepted by industry than management by objectives (MBO). The Farm School has been endeavoring to implement this system by having department heads prepare their annual objectives after reviewing those of the division head, which are based on the objectives of the associate director and the director. Objectives should include (1) definition of the objective, (2) responsibility for implementation, (3) completion date, (4) budget provisions, and (5) outside support required. Implied in the concept of management by objectives is self-control. Managers are expected to inform their supervisors when they are not meeting their objectives. At the same time, however, associate directors are responsible for ensuring that each of their departments is meeting its objectives.

Implementation of management by objectives at the Farm School has given rise, particularly among department heads, to a number of problems inherent in a developing society. One obstacle arises from a lack of discipline among staff members, who complain that they do not have sufficient time to complete the aspect of the program for which they are responsible. However, falling behind schedule often results from mistaken priorities or fear of added responsibility. Considerable effort is required to train subordinate staff to understand the concept of management by objectives, which is the antithesis of traditional peasant management practice. Greek employees are convinced that every laborer or clerk needs rigid supervision and firm discipline, although a more liberal approach is gradually evolving.

SHORT-TERM PLANNING

Both overoptimism and an inclination to avoid unpleasant tasks hinder effective short-term planning in most institutions. Farm School staff members are constantly interrupted in their work by a stream of unscheduled visitors requiring time and attention. Certain techniques have been helpful in overcoming these problems.

An effective device for day-to-day planning is the 3″ × 5″ card. The story was told to a Farm School administrator of a man who made a proposal to the president of a big steel company. He requested half an hour for consultation from each of the ten top executives. He stipulated that he wanted no payment unless the efficiency of the company had increased sufficiently at the end of the year to warrant a $25,000 fee. A year later he received the bank check. His proposal had been quite simple. He had asked each executive to make a list every morning on a 3″ × 5″ card, similar to those used in library reference trays, of the five most important jobs for that day. Regardless of what else they did, they committed themselves to accomplishing the items listed on the card.

This approach was introduced to the top staff at the Farm School, who accepted it with reservations. Only by observing the increased effectiveness of the managers who did adopt this system were the others convinced of its value. Over the years the 3″ × 5″ card has become a mark of an effective manager, although some use variations of the same principle. The staff members have identified three main advantages to this system: The card is readily available to write down an idea; it commits the user in writing to dealing with his problems; and it provides a means of self-evaluation at the end of each day.

The school staff members also use annual pocket or desk calendars to assist in their planning process. Anyone who has grown up in a Western country may be surprised that this even deserves mention. However, the use of planning calendars is comparatively new to Greece and is doubtless a strange idea to training center personnel in Third World countries. A loose-leaf variety, which has been used at the school for many years, has interchangeable pages, which allow permanent records to be kept for several years. The reduction element of copying machines, which can reduce the size of financial reports and other materials, enables managers to carry budgets, student lists, departmental objectives, and other useful information with them at all times.

To ensure effective communication the school circulates a monthly as well as an annual calendar. The monthly calendars indicate dates and times for activities such as board meetings, staff meetings, short-course programs, and daily schedules. To prepare these calendars, each department must anticipate its schedule of activities for the year and participate in joint planning and coordination among the departments. These schedules tie in closely with the statement of the overall objectives of the school's two main divisions: education and administration-production. They provide everyone in the institution with a clear understanding of the goals for the year and the plans for their implementation. By looking at these calendars each staff member involved in a variety of activities is aware of the previous commitments of other staff members. "Avoid surprises" is wise counsel for any institutional administrator.

Although the description of short-term planning at the Farm School indicates that the institution is very efficient and effective, this has not always been the case. However, the guidelines do exist, and constant effort is made to encourage everyone to use the 3″ × 5″ cards, the pocket diaries, and the annual calendar. The staff members have learned over the years how important flexibility is in planning. When they were originally asked to prepare one-year plans the almost universal reaction was, "How do I know what will be happening in six months' time?" They have found, however, that it is better to plan and even plan wrong than not to plan at all.

COMMUNITY DEVELOPMENT PLANNING

The Thessaloniki Community Development Program, which was organized under the capable leadership of Dr. A. E. Trimis in cooperation with the Thessaloniki prefect and his staff in 1958 was one of the school's most successful ventures, and it demonstrated the peasants' ability to plan.[3] It grew out of the recognition that the villagers could do many things for themselves rather than wait for outside leadership and assistance.

Community Development Committees

The first step in the program was to organize community development committees in each village. These committees consisted of elected and informal leaders who were designated by the villagers themselves. Soon after the program began they discovered that no women had been included in the committees—an oversight that was quickly corrected. In the meetings with these committees, the village leaders made it clear that they wanted to know what other villagers were doing about planning and that they needed the support and advice of local government officials, who often appeared to interfere with their plans. They felt that without this cooperation they could do little.

The next step was to organize a voluntary prefecture-level community development committee made up of key officials associated with the villages, including the directors of public works, agriculture, the agricultural bank, the veterinary service, welfare, tourism, and education and the local bishop. This group spent considerable time planning a three-day conference for fifteen to twenty leaders from each of six villages. The speakers at the conference who were members of the development committee were given fifteen minutes to speak on the question, "What can you do to develop your village in my specialty, and how can we help you?"

The fifteen-minute presentation was followed by a time for questions to make sure that the speakers had not been too theoretical. After each group of speakers in one area of interest, such as agriculture, education, or public works, had finished, the gathering broke up into village groups, which discussed the projects presented by the officials or other villagers. During these group discussions, they decided which projects were appropriate for their village and when they could be implemented. They tended to be overoptimistic about what they might accomplish because of their enthusiasm.

On the final day the representatives of each village were asked to report to the conference on their plans for the coming year—a very exciting moment for all of them. Even though they obviously could not accomplish everything, they were encouraged to start on small projects. They realized that they would have a difficult time when they returned home because their fellow villagers would not initially share their enthusiasm. They were encouraged to set up a similar conference at the village level, in which each of the conferees would take the role of one of the speakers to convey their enthusiasm to the villagers.

Many villagers were convinced that nothing would be done. Although some of the speakers were pessimistic, they were willing to try. At the end of the first year the Prefecture Committee discovered that although the villagers had not carried out all the programs to which they were committed, they had accomplished a great deal and had established communication among themselves and with the officials. Most committees had met once a month as agreed. The civil servants and the villagers had changed their attitude toward each other. When extra money was needed from a government source the officials involved, who had developed a much closer tie with these villagers, were somehow able to find it. Even such complex tasks as new roads, water supply systems, and new school houses were completed.

At the same time as this program was being organized at the local level, an effort was made by the National Foundation and the Ministry of the Interior to form similar community development committees in each prefecture and at the national level. About ten years later a government that believed in the authoritarian approach to solving village problems came into power and appointed its own officials, who ended the program. Until then many of the villages continued their planning meetings and implemented a variety of projects.

Twenty-five years after the Thessaloniki Community Development Program was started, village leaders continue to speak enthusiastically about it. Roads, churches, water works, and school houses built during that time are visible proof of its results. The trees planted then have grown into small community forests. More important, many of the leaders of that time learned to work together for the benefit of the community. "Is peasant planning possible?" the Farm School was asked when it started its program. Peasants, community leaders, and government officials working together proved that it definitely is.

Farm School Alumni

Another group that has demonstrated a thorough understanding of the planning process is Farm School alumni. The school received a grant from the Rockefeller Foundation to determine if an institution like the Farm School could introduce a new technology, such as raising broilers, by working through its students and alumni. At a time when chicken was considered a great luxury, a group of twenty-five graduates was invited to the school for a short course in broiler raising. They were provided with a bill of materials from which to build a poultry house, shown how to build it, and trained in feeding, managing, and marketing. They built their poultry houses and purchased the chicks and feed, which the school produced with a loan from the Agricultural Bank. Within three years the whole structure of the Greek poultry business was changed because of these graduates and other producers in another area. As chicken became the cheapest meat, this new group of young entrepreneurs who had learned to prepare long-term plans became established businessmen.

A large number of Farm School graduates from poor peasant families have become master farmers. Some of the top pig operations in Greece with more than seven thousand feeder pigs are run by graduates who knew little when they came to the school. Others are producing early vegetables in plastic greenhouses, and another group operates an extensive network of farm machinery repair shops. Many cultivate their own fields and rent large areas from other farmers to make full use of their high-priced farm machinery. Planning has played a major role in the success of these graduates.

AIDS TO INDIVIDUAL PLANNING

Dr. Harry Peirce, who spent three years at the Farm School as a consultant developing the concept of teaching packages, introduced a new approach to help farm families plan more effectively. His method

SAMPLE GOALS WORKSHEET

Goals	Family Goals	Income Needed (in drachmas, drs.)	Production Goals (area in stremmas, str. — 1,000 m²)	Changes Needed in Production Methods
Long-Term (5 to 8 years)	1. Higher education for children 2. Dowry for daughters 3. Starts for sons in farming or other business 4. Comfortable home 5. Retirement income	300,000 drs. (annual net income) 10,000,000 drs. (net worth)	1 ton corn/str. 400 kilos wheat/str. 250 kilos tobacco/str. 8 tons tomatoes/str. 50 kilos olives/tree 5,000 kilos milk/cow	1. Expand farm business by buying additional land, leasing additional land, and changing to labor-intensive enterprises 2. Mechanize operation to save labor
Intermediate-Term (3 to 5 years)	1. New tractor 2. New car 3. Addition to house 4. Trip abroad	230,000 drs. (annual net income) 6,000,000 drs. (net worth)	Between short-term and long-term	1. Soil test all fields 2. Irrigate if appropriate 3. Buy improved cattle 4. Expand dairy facilities 5. Buy improved machinery
Short Term (1 Year)	1. Good clothes for family 2. New furniture and appliances 3. Vacation trip 4. Better used car	170,000 drs. (annual net income) 3,000,000 drs. (net worth)	10%-20% higher than present yields	1. Keep farm records and analyze 2. Plan crop and livestock program 3. Soil test grain fields and fertilizer 4. Use artificial insemination 5. Use improved seed 6. Use herbicides and insecticides

involves adapting farm practices to changing family goals. The process applies to short-, medium-, and long-term goals. After reviewing the crops that have been produced during the previous year, families are encouraged to set their goals for the coming years. They can see that in order to increase their income to meet these additional obligations they must change cropping plans or find other sources of income. In this way their farming plans grow out of their anticipated needs as they have defined them. Although the school has not yet had a chance to test this approach fully, discussions with agriculturists and farmers indicated a great deal of interest and enthusiasm. A sample work sheet developed by Dr. Peirce indicated the areas to be considered.

Thirty-five years ago it seemed doubtful to most Greek people that either the peasant or the civil servant assigned to help him would be able to make long-term plans for their farms, their lives, or their communities. Yet growing numbers of peasants have learned to plan, and at the same time rural development workers have become more able to help them in their planning. Surely this progress must inspire peasants and civil servants elsewhere to follow in their footsteps. "Make no little plans," said Daniel H. Burnham, a nineteenth-century architect, "for they have no magic to stir men's blood and probably themselves will not be realized. Make big plans, aim high in hope and work, remembering that a noble, logical diagram once recorded will never die, but long after we are gone will be a living thing, asserting itself with ever-growing insistency."[4]

Most peasants through the ages anticipated little more for their children than what they had themselves. But some peasants had aspirations for their children and their grandchildren. The success of their dreams was assured because they were based on plans that would turn them into reality. The kakomiris, the loser, formulated unrealistic plans, whereas the nikokiris included specific objectives with time frames, clarified responsibilities, estimated costs, and anticipated results.

Most farmers are at a serious disadvantage compared to their urban brothers, who are able to close the shutter on their store or the door to their office at night and not worry about the weather and other factors beyond their control. No matter how carefully the peasant plans he still must say, like Hodja, "This is my plan. I think it will succeed, if God is willing." But the winner among the peasants is the one who makes use of every available resource to ensure that as far as his efforts are concerned, his plan will succeed.

6

Organize the Organizers

- How can development workers be trained to help others to organize their work?
- Is organization too Western a concept to be accepted in a peasant society?
- How can the untapped potential of leadership among women be incorporated in community organization?

One day two small boys decided to play a trick on the Hodja. With a tiny bird cupped in their hands they would ask him whether it was alive or dead. If he said it was alive they would crush it to show him he was wrong. If he said it was dead they would let it fly away and still fool him. When they found the wise old man they said, "Hodja, that which we are holding, is it alive or dead?" Hodja thought for a moment and replied, "Ah, my young friends, that is in your hands."

In rural development, organizing involves coordinating the factors of production—labor, fields, equipment, supplies, savings, or credit—to assist the peasant in the management process. For development workers and peasants to understand the rudiments of organization, they must first learn to recognize what is "in their hands" and be trained how to use it. In response to a question about what makes a farmer or village businessman successful a shrewd stonemason replied, "It depends on the person's capacity to organize." This response applies equally to those who manage development programs in rural areas.

Orderliness is one of the most obvious indicators of organization. Farm machinery properly stored out of season, well-maintained buildings, groomed livestock, and neat vegetable gardens demonstrate the presence of a master farmer or institutional manager who has organized his time, facilities, and labor for maximum productivity. It is possible to find unkempt farms or institutions that are productive, but they tend to be exceptions.

Orderliness can also be a measure of organization in an institution as indicated by a comparison of two hospitals in very different settings: the first on the edge of a jungle in Tanzania and the other in a far more advanced country. The first hospital was outstanding from the point of view of organization. The Sisters of the German Catholic order who operated the jungle hospital were radiant, their clothing neat and crisp, and every piece of equipment was scrubbed until it shone. Patients responded to the warmth and compassion that pervaded the atmosphere. The orderliness of the hospital was impressive in contrast to its surroundings in which lepers begged, termites tunneled in the fields, and wild animals prowled the nearby jungle.

The atmosphere was so different from that of a hospital in the relatively developed country, in which the nurses' smocks looked sloppy, no one seemed to care much about the patients, the buildings needed painting, and the corridors were dirty. A visitor commented that patients were sometimes in worse health when they left the hospital than on their arrival.

A similar disparity can be seen between the farms in Holland and those in Southern Europe. On a typical Dutch farm the animals, the barns, and the fields are perfectly cared for, giving a sense of immaculate order. Visitors who ask Dutch farmers why everyone keeps his farm so tidy are told that if a man will not look after his farm the neighbors see that he does. It will take some years before peasants in developing countries catch up with their Dutch counterparts. In Greece the well-organized farmer who operated an orderly farm was in a minority and therefore not in a position to pressure his neighbors to follow his example.

The disparity between the two groups of farms is not only the result of the difference in wealth. Well-organized institutions give the impression of making optimum use of available resources. Once plans and objectives have been defined the responsibility rests with management to establish the organizational structure and procedures to attain the desired goals. Even the smallest training centers must develop an administrative chart that clarifies staff relationships and a plan of work that helps people understand what is expected of them. Capital and operating budgets along with related inventories and replacement programs provide accepted guidelines.

The implementation of a school's objectives by means of its organizational structure depends very much on the individuals involved. Each employee must accept responsibility for giving a full day's work for a full day's pay—a concept not always acknowledged by the institution's employees. The plans must be regularly assessed to ensure that all aspects of the school's work are integrated. Clear communication among the staff members is essential, both vertically and horizontally, to create a sense of team work. David Willis, associate director for administration at the Farm School, described this process: "Our job is to plan our work and work our plan."

ORGANIZATIONAL GUIDELINES

A few guidelines that have been implemented in the Farm School may be useful to development workers elsewhere. The school's administrative chart will help the reader understand its organization. The function of such a chart is to define the working relationships among the staff to ensure that everyone is making a maximum contribution toward the school's objectives.

As can be seen in the chart each employee reports to only one supervisor. It is important to clarify the unity of command to make sure that only one individual in each department has the authority to make decisions. The number of employees that reports to any one supervisor should be limited. Although this number depends on the activities of a department, the ability of the workers, and the geographical area covered by the program, no one person should have more than six to eight individuals directly responsible to him. Closely tied to the interrelationships among the personnel is the type of cooperation required among volunteer workers, trustees, trainees, and staff. Programs in which trainees and students are involved in decisionmaking are almost always more successful. This also applies to participation by technicians and laborers in planning and organizing within their own department.

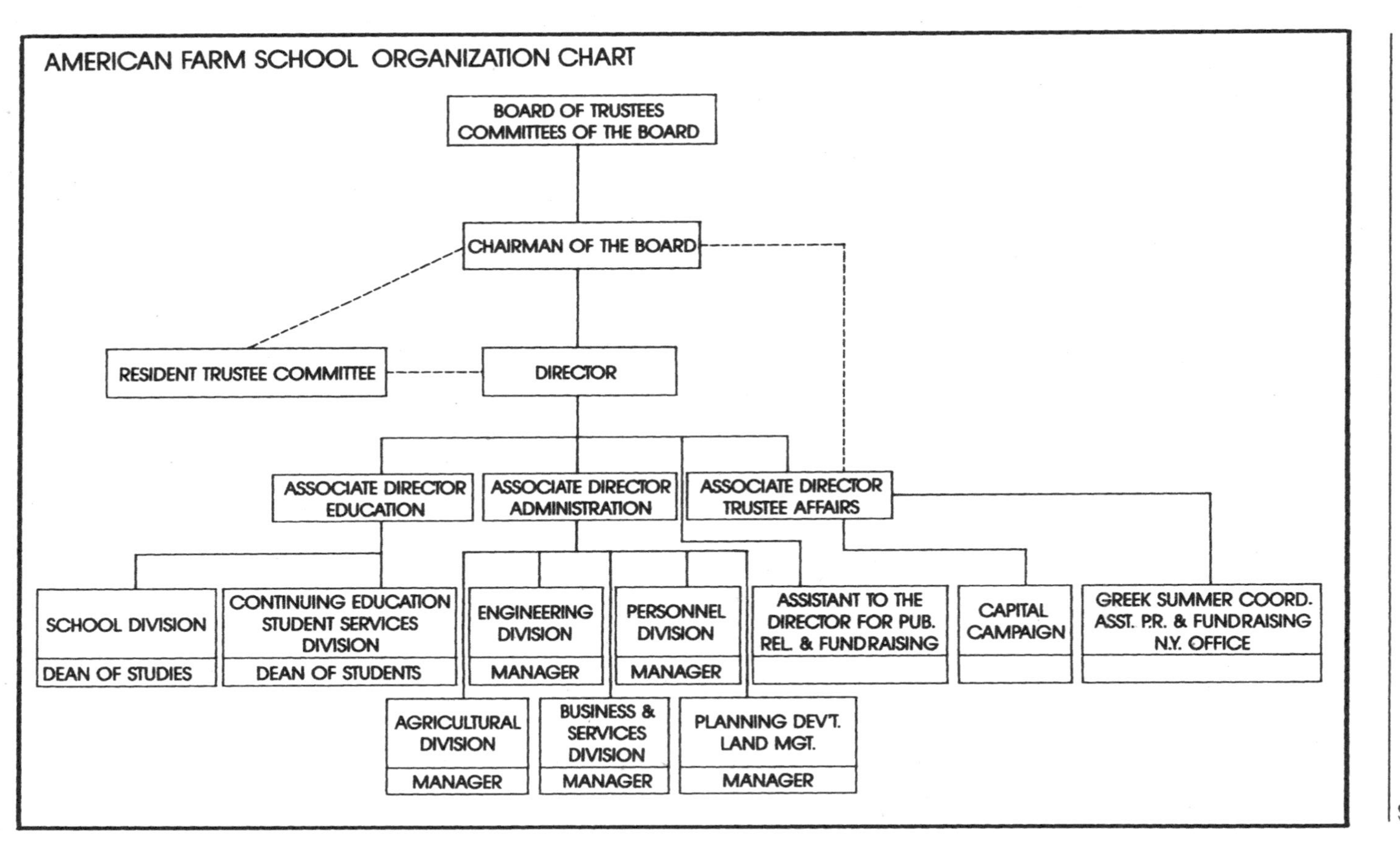
AMERICAN FARM SCHOOL ORGANIZATION CHART
BOARD OF TRUSTEES
COMMITTEES OF THE BOARD
CHAIRMAN OF THE BOARD
RESIDENT TRUSTEE COMMITTEE
DIRECTOR
ASSOCIATE DIRECTOR
EDUCATION
ASSOCIATE DIRECTOR
ADMINISTRATION
ASSOCIATE DIRECTOR
TRUSTEE AFFAIRS
SCHOOL DIVISION
DEAN OF STUDIES
CONTINUING EDUCATION
STUDENT SERVICES
DIVISION
DEAN OF STUDENTS
ENGINEERING
DIVISION
MANAGER
PERSONNEL
DIVISION
MANAGER
ASSISTANT TO THE
DIRECTOR FOR PUB.
REL. & FUNDRAISING
CAPITAL
CAMPAIGN
GREEK SUMMER COORD.
ASST. P.R. & FUNDRAISING
N.Y. OFFICE
AGRICULTURAL
DIVISION
MANAGER
BUSINESS &
SERVICES
DIVISION
MANAGER
PLANNING DEV'T.
LAND MGT.
MANAGER

Volunteers must be discouraged from interfering with the internal lines of communication established by the organization chart.

Departments can be organized according to their function or their product. The supervisor in a line function directs others and is responsible for a set of operations or activities such as instruction or production; those with a staff function do not issue orders except to their own assistants but do guide and advise line supervisors on such matters as finance and personnel. When supervision is mixed (as indicated by dotted lines in the chart), it is even more important to have a clear definition of relations and responsibilities in order to avoid conflict. Failure to make these clarifications in advance is one of the major sources of friction in most institutions.

Various other tools are helpful to a director in organizing a program. Two that have been particularly useful at the Farm School are a personnel manual, which states the rights and responsibilities of both the employee and the employer, and a set of job descriptions, which defines the tasks expected of each employee and his relationship to others.

It is extremely important for development supervisors to learn to delegate their authority and responsibility to subordinates so that others are in a position to deputize for them when they are called away. Delegation is also good training for understudies, who are sometimes reluctant to accept the responsibility passed to them by their supervisor. In the management of a family farm, one of the great difficulties is to encourage parents to give authority and responsibility to their children. This participation would not only involve the young people in decisionmaking but also give them a feeling of importance in the farm enterprise. Many Farm School graduates have been extremely frustrated by the unwillingness of their parents either to accept their innovations or to allow them to participate in management decisions.

ORGANIZATIONAL FUNCTIONS OF MANAGERS

A fundamental question for a development worker is how to help the peasant organize his operations more effectively. The Farm School has identified a number of basic organizational functions that are useful in training peasants to become better managers.

One of the most important basic techniques is budgeting, a process seldom used by peasants. The Greek government has begun to gather records from selected farmers and to feed them into a central computer. Specially trained extension agents work with a small group of farm families helping them collect data on yields, size of operations, number and type of livestock, and cash income and expenditures. These data provide the basis for comparing Greek agriculture with its counterparts in other Common Market countries. These records are very useful for the Greek Extension Service in training farm managers. For many years the primary concern of the Extension Service was production; only recently has more emphasis been placed on farm management and budgeting. At the Farm School essential aspects of the student projects are budgeting and record keeping; skill in these areas is required as an integral part of their profit-making businesses.

A closely related aspect of effective organization is good credit planning. For centuries Greek peasants have been at the mercy of the merchants, who loaned them money at usurious rates. Peasants must become familiar with the concepts of debt, interest costs, and rates of return. The organization of the Agricultural Bank in 1923 and subsequent legislation on the formation of cooperatives have played a fundamental role in providing adequate credit for farmers. The Agricultural Bank, which is a nonprofit institution, maintains two separate departments: one for financial administration and the other for technical advice and training. Before a farmer is granted a loan his application is reviewed by the technical staff of the bank, who examines his operations and planning to ensure that they are feasible. The bank makes available short-, middle-, and long-term loans for agricultural enterprises and private and cooperative processing facilities in the villages, thus encouraging the peasants to stay rather than move to the cities.

To achieve the goals established by the family, a good master farmer integrates the organization of his activities with his planning. A visit to almost any village in Greece reveals half-finished farm structures, begun as part of a building plan that did not adequately consider the limitations of credit, cash flow, and market forces. In many cases, farmers with spare cash from a successful crop year invest it in the concrete work of a new building, as a safeguard against inflation, fully aware that they will not have enough money to complete the project. Some shrewd villagers built poultry houses on low-interest credit from the Agricultural Bank and subsequently converted them into lucrative tourist apartments.

The organization of marketing is one of the weakest management links in a peasant economy. The main reason for this shortcoming is the peasant's assumption that marketing is someone else's business—a result of living in a subsistence economy. Many Farm School graduates have realized that it is far more profitable to organize their own marketing than to depend on merchants. Some have established such a good reputation that they are able to market their neighbor's products as well as their own, both within Greece and abroad.

Machinery management, relatively new to Greece, is a major factor in the success of an operation. Peasants buy big equipment because they think it will be more profitable or because it gives them added status. One Farm School graduate who owned seven acres of land inherited another seven from his father and has since bought an additional seven acres, making the area of his farm three times the national average. Surprisingly, he does not own a single piece of machinery, even though he is recognized as a successful farmer. Because irrigation was impossible on the dry land that he owned, he could grow only wheat, barley, or unirrigated cotton. So many tractors were available in his village that he was able to hire them to work his fields. He was only required to supervise the farming, freeing him to work as a plumber. Other graduates have chosen to invest their available resources in farm machinery and to cultivate their neighbors' fields on a contractual basis. In many cases they have become highly successful entrepreneurs and leaders.

Diversifying farm operations in Greece to reduce the dependence on one crop has not always been possible. Experts have spent years trying to find a replacement for tobacco on the infertile Greek hillsides. In some areas graduates of the school have discovered that vineyards producing wine and table grapes for export are a good substitute. Selecting crops that distribute the demand for labor evenly throughout the year is equally important. Diversification also spreads the risks from weather and market oversupply. Since the formation of the Agricultural Bank farmers have been able to purchase crop insurance to protect them from hailstorms and winter freezing. One graduate with a heavy investment in combines and other equipment increased his returns with crop insurance. Once he received payment on his insurance from damage to winter grains after a severe freeze, he planted a spring crop that yielded a second income from the same field in one year.

Farmers must develop an organizational structure even for family operations. Some of the school's graduates have designed operations using only family labor, to make maximum use of human resources throughout the year. Tobacco-farming families, who have moved to the city to put their children in school, return to the village for three months of intensive labor every summer, after which the father remains to market the crop.

Public relations is an important aspect of even a small farming operation. Farmers are realizing how dependent they are on good relations with the neighbors, their employees, the Agricultural Bank, and the Extension Service. Although farmers once tended to be hostile and suspicious of these groups, particularly in financial dealings, they have become more outgoing and friendly. Over the years Farm School graduates have acquired a reputation for maintaining good relationships with business associates, neighbors, and government officials.

Finally, peasants must learn to evaluate and innovate constantly. They must keep track of what others are doing, both in their own as well as other occupations related to agriculture. They should ascertain whether they can obtain a larger share of the business by expanding their activities to include marketing and supply as well as production. They must inform themselves of current trends in more advanced countries to determine where their organizational emphasis should be placed in the future.

RELATING TRAINING TO PRACTICE

It is very important for an agricultural institution to train its students for as many aspects of their future life as possible. It is not sufficient to concentrate on improved agricultural techniques or even just management; training must relate to the variety of challenges that they will face back in the village as agricultural technicians and managers and as members of a family. The Warrentown Agricultural School in Ireland has organized an annual Farm Management Competition for students of the senior class in which they can use their training in a practical situation. The class is divided into teams of four, who visit a neighboring farm operated by a master farmer. The students make a complete study of the farm and are allowed three days in which to ask questions and collect all the information they need. They then have a period of approximately two months to prepare a plan and organization for its implementation on the farm. At the end of this period the students submit their findings to a committee of teachers, local extension agents, and the farmer himself, which awards a prize for the best plan. This prize is even more coveted than athletic trophies.

A useful approach to training young people has been implemented at Donart High School in Stillwater, Oklahoma. The school offers a one-term course for fifteen boys and fifteen girls, who work through a series of imaginary life situations as couples. They develop a practical understanding of the planning and organizational elements involved in getting married, buying or renting a house, and starting a family. The first part of the course covers the first year of marriage, complete with a baby represented by a raw egg, which is constantly in the possession of the boy or the girl. They also discuss topics relating to insurance, taxes, credit, and banking. For the subsequent period they cover such decisions as staying on the farm or moving to a city and long-term family planning. The final lesson in the course is burying one's spouse.

The course was based on a kit by Cliff Allen called Practical Family Life, which provides an excellent collection of pamphlets for easy reference. When this idea was introduced to the faculty of the Farm School they felt that it was too daring, although a few were convinced that, properly administered, the course would be extremely useful for both high school level students and young married couples in the school's continuing education programs.

A most successful technique was used by the Farm School and the Extension Service to develop home poultry flocks in several villages. About twenty-five men and women from a key village were invited to the school and given in-depth training on setting up a home poultry flock. This training included every aspect of the operation, with primary emphasis on organization. They were given the design of a poultry house, a bill of materials, and an opportunity to purchase baby chicks from parent stock raised at the school. They learned about the need for different feeds at various stages of development as well as the elementary aspects of management and disease control. The villagers went home and established their home flocks with great enthusiasm and optimism—the first step in the Farm School program.

Subsequently other villages were selected to follow the same pattern, beginning with three-day short courses. On the first day the trainees came to the school and learned the basic principles of village poultry raising. They spent the second day in the key village talking to some of the twenty-five people who were now operating their home poultry flocks. These meetings gave them an opportunity to evaluate the program and to hear from the villagers themselves how they went about organizing the operations and what profit they had made. On the third day, in a session at the school the trainees were full of questions about setting up their own home flock. When they returned to their villages and implemented what they had learned, their poultry operation served as the example for their neighbors. Eventually the villagers in the key village began incubating hatching eggs to sell chicks to the other villages.

Many Farm School students have been attracted to the Farm School because a specific area of expertise at the school is important to their career in farming. Such students usually become the most successful graduates because they start applying what they learn at the school back at the family enterprise even before they graduate. Athanasi Karakolis, who was anxious to help his father in the turkey business, came to the Farm School because it produced and marketed the best turkeys in Greece at premium prices. His father was eager for him to have both theoretical and practical training in management in the production units. The boy acquired additional experience in Holland during the summer as an apprentice to a Dutch farmer.

WHO IS THE MANAGER?

One element of organization that deserves special consideration in the management of the family enterprise arises from the confusion among development workers about who is in charge of the farm family. The Greek village story contradicts the generally accepted understanding that the husband manages the peasant family. Peasants often say, "The husband is the head, but it is the woman who is the neck and decides which way the head should turn."

There is a clear distinction between role playing and reality. During the years of the Turkish occupation (before 1912 in Northern Greece) women tended to be considerably younger than their husbands and to have far less contact with the outside world, which made them very dependent on their husbands both economically and socially. Even as they gained independence and self-assurance they were expected to behave in a subservient manner to their husbands in public, regardless of who was actually the dominant partner. The man who married a rich bride and lived with her family was the constant butt of jokes because he was regarded as having condemned himself to a life of servitude.[1]

However, in the typical family, which lived with the husband's parents, the mother-in-law, not the husband, wielded the power and made sure that the bride waited on her son hand and foot. The bride had her own subtle ways of counteracting the mother-in-law's authority by using "bed murmurings" (translated as "pillow talk"), which in Greece are generally recognized as a one-way conversation always from wife to husband.

A village priest explained the transition in the status of women: "Our mother and the children were afraid of our father. He used to beat her and all of us quite regularly, but in my generation my wife and children respect me." He did admit to beating his daughter for having had an illicit affair, following which she eloped. "In her generation," he continued, "my daughter makes all the decisions in the family, but that is because she is more intelligent."

Status and face saving are strong motivating forces among the villagers as are changing economic factors. George Skouras from a refugee village observed, "The woman decides everything nowadays, even what the man is going to plant in the fields. With no available labor the husband cannot pick the cotton or hoe the melons without her, so that she has to agree in advance on how much land they can work."

Most analyses of decisionmaking indicate that the wife makes the decisions about food, the children (except when the mother wants to use the father as an authority figure), the vegetable garden, and domesticated animals. She deals with the traveling vendors and decides when the children need clothes and what items must be bought for dowries. House furnishings and essential appliances are also the woman's responsibility. Such decisions as crops to be cultivated and acreage, type

A village story tells of a great argument between two men about who runs the family. The first one offered a rooster to every household that was run by the woman. The second bet a horse for every household run by the man. At every house they asked the housewife who ran her family. Whether it was a young bride, a grandmother in her seventies, or a middle-aged mother, each admitted to making the decisions. At the edge of the village they saw a young handsome man standing in front of his door and asked him who ran his house. When he said he did, the second man offered him one of the horses and asked him if he would like a brown one, a white one, or a grey one. The young man thought for a moment and replied, "Just a minute, let me ask my wife what she thinks."

and size of farm machinery, the choice between a tractor and a pick-up truck are usually made by the man, but the woman has a veto and often uses it.

Husbands admit to being hesitant about making decisions without consulting their wives. The wife may not appear to make the decision, but because he dreads criticism, shame, and conflict the husband ascertains his wife's views. Christo, a village builder, gave the following appraisal of family relationships: "Without my wife I could do nothing. She keeps the 'money pouch' and decides on expenditures. She also handles tax matters. When we have special building projects she even comes along and helps me." Their daughter, a first-year university student, says that her mother makes most domestic decisions, but major issues, such as where they will live or whether the daughter would go to the university, are decided by the family jointly.

Originally, cash in the village family came from the sale of eggs or pieces of embroidery by the wife, so that she normally handled actual money. Today she does the budgeting and record keeping, although on a rather limited scale. As the banker the woman has considerable say in all purchasing, financing, marketing, and other fiscal decisions. Some families also have large sums, perhaps equivalent to US$5,000, hidden in secret corners of their houses for emergencies. A story was told by Joice Loch, an old-time resident of Greece, about the village family that came to her one day begging for help. They held in their hands the shreds of thousands of drachma notes eaten by mice, which they thought she could turn into the bank for new ones.

A study should be made about the growing influence of older children in Greek families. Parents report that university or junior college students now play a dominant role by telling them how things are done by "civilized" people. The parents appear to be apprehensive about the imaginary city person or potential bride or groom who might discover how "primitive" they are.

IMPLICATIONS FOR DEVELOPMENT

Emphasis must be placed on training both men and women, boys and girls, in every aspect of rural life, beginning in secondary schools and carried through to adult short courses. It is not the function of innovative education to wait for changing values but to prepare for them. Particularly vital is joint training for husbands and wives; if they are to decide together, men and women must learn together.

Specialists agree that much more needs to be done in joint management training for men and women. Anastasia Xenou of Greece's Common Market Division of the Ministry of Agriculture, insists that the woman really is the decision maker in the rural family, and yet no training in management is available for her at the elementary, secondary, or postsecondary level to prepare her for this role. Professor Clio Presvelou, chairman of the Department of Home Economics at Wageningen in Holland, speaks of the importance of working with the family as a unit,[2] a nucleus recognized by Aristotle when he referred to economics having its origin in the home.

A working committee of educators, home economists, and specialists from the Greek government has helped to design a course in rural economics and management, which will include case studies in managing the home, borrowing money, buying or building a home, budgeting, deciding whether to stay in the village or move to the city, insurance, taxes, social security, money management, purchasing a car or tractor, long-range planning, and operating a rural business such as a grocery, hotel, or restaurant, as well as elective courses in home economics, crafts, and recreation. Clearly more life preparation courses need to be organized for village boys and girls together.

In a study on the future directions of home economics in Common Market countries, Nomiki Tsoukala, the director of home economics in Greece, said, "Women must be just as well informed as the farmers to improve their agricultural work and the quality of their products."[3]

Development workers must find ways to operate programs based on peasants' needs, to recognize that they must be for the whole family, and to encourage men and women to resolve their problems together. The terms *peasant, villager, master farmer,* and *manager* are used throughout this book. They refer equally to women and men—possibly even more to women in the area of management.

The contribution of effective organization to good management is obvious in both a development program and the peasant enterprise. A number of useful guidelines to proper organization have been presented for both, along with some suggestions on practical methods of teaching organization. A significant shortcoming in management training in the past has grown out of a failure to understand the organizational structure of the peasant enterprise, with particular reference to the important role of woman in joint decisionmaking. Management training for villagers must begin with the premise that just as men and women live, work, and decide together so they must be trained together if a development program is to be successful.

7

The Languages of Leadership

- How is peasant leadership related to management training?
- Is there a special language that leaders must use?
- Can leadership be learned by peasants?
- What are the qualities they must develop?

For many years there has been a bronze casting at the Farm School of Hodja riding on a donkey. It relates to the time when he was seen in the village riding his beloved donkey backwards. When a neighbor asked him why he was facing that way Hodja said, "My friend here wanted to go one way and I wanted to go the other, so we are compromising."

Leadership is the quality that ensures that a manager, his or her associates, and the organization are all moving in the same direction to the same destination at the same speed. It is as important an attribute for the development worker, who must gain the respect and recognition of the villagers, as it is for peasants.

The Greek language has two words for "leader": *Ygetis* means a natural leader, and *archigos* refers to an appointed leader. The difference in their meanings became clear when a group of youth club leaders attending a seminar on leadership were asked to name some of the great leaders in history. Their list of names included

Churchill	Hitler	Pasteur
Karamanlis	Alexander the Great	Roosevelt
Napoleon	Venizelos	MacArthur

The seminar members, in reviewing the list, recognized that Churchill, Karamanlis, Venizelos, Pasteur, and Roosevelt gained their positions of leadership by popular recognition, based on certain innate or acquired qualities. Each was an ygetis, a natural leader. The authoritarian leaders—Napoleon, Hitler, Alexander the Great, and MacArthur—either were appointed or took leadership positions by force. Some appointed leaders may become successful natural leaders because of their personal attributes, even though they originally gained their position through force.

The more the appointed leader of a development program becomes a natural leader the more successful the program will be. This transition raises two questions: How can appointed leaders of development programs be trained to become competent natural leaders? How can peasants be trained to become more effective leaders, particularly in the cooperative movement?

The YMCA has devoted many decades to training both professional and volunteer leaders. Jerold Panas, a YMCA consultant who has spent many years working with volunteers and professionals in fund raising and campaign management, summarized his observations on leadership as follows:

> A leader is willing to take the risk, the blame, the brunt and the storm.
>
> He has the power to persuade and inspire others to heights they thought unattainable.
>
> A leader has the heart and the mind to make decisions quickly and decisively. There is a scourging honesty to all he does, a talent to cooperate and coordinate. When he speaks, others listen.
>
> He ignites sparks which propel others to action.

> When Aeschines spoke, they said: "How well he speaks. What glorious words." But when Demosthenes spoke, they shouted: "Let us march against Philip, now." Such is the quality of leadership.[1]

Describing leadership and its role in development is difficult because of its manifold qualities. Two individuals who embodied many of these and have had an inestimable impact on the Farm School's growth are Theo Litsas and Avrilia Vlachou. Litsas, who was associate director of the school, played an especially dynamic role in shaping its course. Avrilia, who later became Sister Lia, a Greek Orthodox nun, helped to found the girls' school shortly after World War II. She taught the students the "five languages of leadership," which Litsas put into practice. The following sections illustrate the contribution of inspired leadership by people such as these to a development program.

LEADERSHIP QUALITIES

Sister Lia spent considerable time in India working with lepers. She had not been able to master any one language because her patients came from different parts of the country. After she had been there two years, an Anglican bishop asked her if she had learned the language of the natives. Embarrassed to admit that she had not, she hesitated for a moment and then said to the bishop, "Oh yes, your Grace. I have learned five languages." He looked surprised as she went on, "the languages of smiling, weeping, touching, listening, and loving." Learning Sister Lia's languages is essential to becoming an effective leader in a training program.

The Language of Smiling

As a development worker, Theo Litsas was never without a laugh and a story. He worked all day and half the night. More than anything, he brought fun and sheer delight in the game of life to those who lived and worked with him at the Farm School. He was the one who taught the tales of Hodja to the staff: Every time a difficult or frustrating moment came, he seemed to have a Hodja story to match the occasion. He recognized how vital a sense of humor is in running an institution, but he also knew that it involved much more than laughter and jokes.

Most development workers are apt to take themselves much too seriously. In their eagerness to accomplish their objectives, they lose their ability to laugh at themselves and the world around them. When new young staff members become frustrated by lack of progress at the Farm School, they are introduced to Harvey Mindess's analysis of a sense of humor, which often helps them understand the mental attitude associated with the language of smiling:

> A cluster of qualities characterizes this peculiar frame of mind: flexibility, in this case an individual's willingness to examine every side of every issue and every side of every side; spontaneity, his ability to leap from one mood or mode of thought to another; unconventionality, his freedom from the values of his time, his place and his profession; shrewdness, his refusal to believe that anyone—least of all himself—is what he seems to be; playfulness, his grasp of life as a game, a tragi-comic game that nobody wins but that does not have to be won to be enjoyed.[2]

Above all, development workers need to cultivate a sense of humor about their work. This may not necessarily come naturally, but reflection on a Hodja story or its equivalent has been helpful to many of them in facing complex problems.

The Language of Weeping

Compassion is an integral quality of leadership. Its role in a training center is particularly important for the director (see discussion in Chapter 4). Trainees come with their own worries, fears, and sometimes tragedies. If the leader does not take time to listen, they will not pay attention to what he wants to tell them. Henri Nouwen, a Dutch Jesuit priest who taught at Yale Divinity School, describes the compassion of a spiritual man:

> Across all barriers of land and language, wealth and poverty, knowledge and ignorance, we are still one, created from the same dust, subject to the same laws and destined for the same end. . . . His flesh is my flesh, his blood is my blood, his pain is my pain and his smile is my smile. There is nothing in me that he would find strange, and there is nothing in him that I would not recognize.[3]

In addition to its programs for Greek young people and adults, the Farm School also operates an international summer program for teenagers to work in a Greek village. Soon after forty campers had arrived from abroad one summer, word reached the school that the brother of one of the boys had been killed in an automobile accident. A small group met to discuss how to help the bereaved brother but felt inadequate because they did not know him very well. One commented, "You have to laugh with someone before you can cry with him." Successful leaders of any program learn to laugh with their trainees so that they can cry with them too.

The Language of Touching

People seldom realize how powerful are the physical aspects of a greeting. Westerners in particular seem reserved in their salutations. The Indians welcome a person with their hands in a position of prayer saying "Namaste"—"I worship the god in you"—and seem to greet the whole being. Greek villagers are unsurpassed in their welcome, throwing their arms around the visitor as if he or she were the most important person in the world. They use the word *angaliazo*, which can variously be translated as "embrace," "squeeze," "press close," "fold in one's arms," "clutch," and "cuddle." Education very often seems to inhibit touching and angalia. Foreign teenagers who live with Greek families never forget the warm embracing of their Greek mother and father and of all the villagers. Theo Litsas made a profound impression on people when he greeted them. He put his whole self into his greeting—his eyes, his smile, and his whole face; his outstretched right hand extended toward the other person while his left reached up to the visitor's shoulder. The Farm School has a tradition of sharing handshakes at special gatherings, which gives students, staff, and guests an opportunity to look into each other's eyes and share the warmth of the other person as a greeting or farewell.

The Language of Listening

Listening intently is as important for a leader as speaking—an attribute of which Theo Litsas was fully conscious, despite his dynamism. He could lead a dance, a song, or a game, but he could also follow and listen—not just briefly but for an hour or an evening—no matter how busy he was. Long after he died, students and friends remembered occasions when Litsas had taken time to become sincerely involved and concerned about a personal problem one of them had discussed with him.

This quality of listening, which won such respect and affection for Litsas, was discussed at an international conference on the role of the volunteer in development. A Greek participant spoke of the three languages in every person: the prosecutor's, which condemns; the parent's or guardian's, which criticizes; and the brother's, which listens and loves. He pointed out that when leaders talk to people with similar backgrounds or levels of education, they are free to use any language they please. But to others they must limit themselves to the language of the brother, which has the characteristic not only of speaking but also of listening and loving. The ability to use the language of listening distinguished Theo Litsas and others like him and is so important for anyone involved in a training program.

Dear Development Worker,

When I ask you to listen to me and you start giving advice, you have not done what I asked.

When I ask you to listen to me and you begin to tell me I shouldn't feel that way, you are trampling on my feelings.

When I ask you to listen to me and you feel you have to do something to solve my problem, you have failed me, strange as it may seem.

Listen! All I asked was that you listen,
not talk or do—just hear me.

And I can do for myself. I am not helpless. Maybe discouraged and faltering, but not helpless.

When you do something for me that I *can* and *need* to do for myself, you contribute to my fear and inadequacy.

But, when you accept as a simple fact that I do feel what I feel, then I can quit trying to convince you and get about the business of understanding what's behind this feeling, even if it is irrational.

And when that's clear, the answers are obvious and I don't need advice.

Thanks for listening.

Gratefully,
Your peasant friend

A capacity for listening was very much alive in an eighty-year-old visitor to the school, Mrs. Georgiana Sibley, who was particularly interested in young people. Long after her visit a student referred to her in a discussion of saintliness. When asked why, he replied simply, "Because she listens when you talk." Trainees regard a development worker who is fully attentive when his trainees or staff members are speaking in the same way that the student considered a person who listened when he talked—as a saint.

The device of a letter (see above) from an imaginary peasant to a development worker can be used to illustrate the meaning of listening.[4] Rather than hearing what a student is saying, the technical assistance specialist tends to feel compelled to act, to bring about change, to see tangible results. This specialist fails to understand how much the individual needs someone who really cares, who demonstrates concern by listening, and who can help the other person actually solve the problem himself.

The Language of Loving

Of Sister Lia's five languages perhaps the most difficult to describe is the language of loving. The development worker's love must be akin to J. P. Morgan's comment on the cost of owning a yacht: "If you have to know how much it costs you, you cannot afford it." The development worker's real challenge is to love himself out of a job. He has to give totally, expecting nothing in return. Some trainees openly convey their appreciation, but he should not be surprised to sense scorn and contempt from others. The real satisfaction for staff members in a training center should grow out of following the progress of the alumni rather than expecting gratitude from them during their training.

One of the best descriptions of the language of loving comes from St. Paul's First Letter to the Corinthians, in which the word *leader* has been substituted for the word *love:*

> [A leader] is patient and kind. [A leader] is not jealous to boastful. [He] is not arrogant or rude. [A leader] does not insist on [his] own way. [He] is not irritable or resentful. [He] does not rejoice at wrong but rejoices in the right. [A good leader] bears all things, believes all things, hopes all things, endures all things. [His work] never ends. (13:4–8)

The Language of Praying

In her later years Sister Lia added a sixth language to her original five, the language of prayer. Her interpretation tied in closely with Dr. John House's motto—"To work is to pray." This spiritual quality gave a special dimension to Litsas' leadership. His Greek Orthodox upbringing influenced by Quaker thought helped him to understand the integral relationship between work and prayer and to interpret it for the students. The Farm School Creed, which describes the commitment to Dr. House's philosophy, is still used on special occasions:

> I believe in a permanent agriculture; a soil that grows richer, rather than poorer, from year to year. I believe in living, not for self but for others, so that future generations may not suffer on account of my farming methods. I believe that tillers of the soil are stewards of the land and will be held accountable for the faithful performance of their trust. I am proud to be a farmer, and will try to be worthy of the name.[5]

VILLAGE LEADERS

Inspired leaders have to learn the practical elements of leadership by experience. Their leadership roles vary according to the needs and interests of each group they direct. They must not only lead the trainees but also manage programs. They may be called upon to supervise other staff members, and they must learn to compromise with workers from other disciplines who are participating in their programs. Depending on the situation, they must be prepared to switch from democratic to authoritarian leadership, and they must sense when to lead and when to follow, being constantly attuned to the needs of their associates and the trainees. This awareness is every bit as essential for development workers as it is for village leaders.

"In every village you may find a few people who lack education, but have within them the wisdom of the ages" was among the last thoughts shared by Charles House with his successor. Effective leaders are those who have sought to emulate the best qualities of their elders, adapting their wisdom to deal with current problems. Identifying and helping to train such individuals are sometimes more rewarding than training development workers because these leaders often have more to share about their personal experiences than they themselves learn. The Farm School staff has made some observations on the characteristics of such master farmers and their wives, based upon their work in leadership training.

There is a false assumption in many cultures that leadership has a charismatic quality that cannot be learned. Leadership training should be one of the primary goals of any development program for extension workers and peasants alike. Under Theo Litsas the Farm School made an important contribution to village development through a variety of courses to train leaders in recreation, community organization, and cooperative management. Even village priests seemed to develop a greater sense of confidence and were eager to try new approaches to gain added recognition in their parishes. Building this self-confidence is a vital element of leadership training.

A farm school staff member working in the villages observed that innovative graduates seldom accepted elective positions in the villages. In contrast, graduates who became village presidents were not generally willing to try out untested agricultural practices. This observation would seem to reinforce leadership studies conducted in the United States, which indicate that there are two distinct types of leaders. On one hand, the innovative leader is willing to experiment even if he fails; he does not have a large following among other farmers because they regard him as eccentric. The group leader, on the other hand, moves ahead of the rest of the peasants but always at a pace that they can follow. Daniel Benor and James Q. Harrison use the phrase "imitable contact farmer" to describe the group leader who makes innovations that the other villagers are likely to follow.[6] In describing such leaders they state,

"But they should not be the community's most progressive farmers who are usually regarded as exceptional, and their neighbors tend not to follow them."

Development workers often make the mistake of overemphasizing the potential leadership qualities in the peasant whose approach seems most compatible with their own, not realizing that he does not necessarily have the confidence of the other villagers. The leader must share the feelings and values of the villagers and will often oppose the suggestions of the development worker, aligning himself with the villagers in their conservatism and unwillingness to change. Choosing someone who constantly agrees with the development worker may seem like the easiest solution in the beginning, but it is bound to end in failure.

Many promising Farm School graduates have not been accepted as leaders in their villages until they have demonstrated their ability as farmers. Some graduates required several years of trial and error before they established themselves as successful producers, and only then were they recognized. During the intervening period the villagers wondered why these young men had wasted four years studying at the Farm School when they did not seem to be different from anyone else in the village.

The village leader, as the spokesperson for the community, must be able to relate to outsiders and speak convincingly. To increase the students' confidence, the Farm School introduced a public-speaking program in which they are required to give five-minute talks to the entire faculty and student body. Long after they have left the school, graduates describe their original stage fright and the confidence they acquired with each successive talk.

Leadership in the village is specific, not general. Some graduates serve on committees for the church or the school; others are more active in agricultural production and become involved in the cooperatives. Very few attain positions of leadership in all aspects of village life. This characteristic of leadership is not limited to the village but applies to all walks of life. Those who become leaders in one sphere are often disappointed when they fail to win support in an unrelated area. It is difficult to help them understand that leadership is associated with a particular group, period of time, and specific place, and it is not usually transferable from one situation to another.

Some Farm School graduates have been appointed by the government to posts such as village secretary, assistant to the agriculturist, or artificial inseminator. Formal appointment to these positions does not necessarily ensure leadership status. They must work hard to win the confidence and respect of others in the village through dedication and the services they themselves provide. For graduates to be recognized as leaders they first have to prove that they are willing to serve the village. They spend considerable time helping particular individuals, listening to their problems, or traveling to the nearest prefecture office on their behalf. They often pay a high price in time and money but feel rewarded by the recognition this approach brings them.

The majority of development workers involved in training peasants are faced with a two-fold challenge: They must broaden themselves to become confident and effective leaders, while identifying and cultivating leadership qualities among those with whom they work. Just as the capable sergeant is continually seeking to improve himself while inspiring others to become more competent, so those working in rural areas must understand that leadership development is an important goal for their own self-enhancement, as well as for the people in their district. Inspired leadership among both development workers and peasants is one of the major contributing factors to rural development.

One day Hodja borrowed a large copper pot that he shortly returned to his neighbor with a smaller pot inside. When Hodja told him that the pot had given birth the neighbor did not really believe him but was delighted with the unexpected gift. Later when Hodja again asked to borrow the pot the neighbor responded with alacrity. After several weeks the neighbor asked what had happened to his pot. Slightly embarrassed, Hodja reported that the pot had died. "Whoever heard of such a silly thing as pots which die?" said the incensed neighbor. "Well," said Hodja, "if there are pots which can give birth, they can certainly die."

8

Maintaining Control

- How does control relate to managing a farm or a development program?
- Can peasants learn to apply control in a village situation?
- What is the role of the budget in ensuring control?

Control is the process of determining whether every part of an organization is adhering to an established plan and progressing toward clearly defined objectives. Using Hodja's logic, control is making sure that neither a neighbor, an employee, nor a competitor can have "pots which die" or others "which can give birth."

INSTITUTIONAL CONTROL

Criton Kanevas, an executive in a prominent Greek company for thirty-five years before he retired to manage a friend's farm, kept careful accounts of each item of the farm's income and expenditure and prepared detailed records of each employee's work schedule. Kanevas described the significance of control in relation to the problem of weeds: "Some people wait until their field is covered with weeds before they start harrowing. Some look for the weeds as they appear and remove them. Others apply pre-emergence weed killer when they plant their seed. Master farmers rotate their crops to stop weeds sprouting in the first place." Control is the process of ensuring that no weeds appear.

Two questions constantly recur in discussions on effective training for master farmers in Third World countries: How can the development worker be certain that he is maintaining adequate control in his organization? How can the peasant be trained to do the same on his farm?

Control is necessary when authority is delegated because ultimately the responsibility still rests with the development worker, who must be able to relate what is being achieved to his original goals. This relationship makes it essential that objectives be stated so clearly that any deviations will be obvious and can be identified early, to ensure that required adjustments are periodically carried out.

These objectives are important for all workers so that both they and their supervisor understand what they are expected to achieve within specified periods. When the concept of management by objectives and self-control is implemented, periodic discussions among a supervisor and the staff members allow the members to report on their work so that the supervisor can evaluate overall progress in relation to the objectives. Performance-based objectives in education (reviewed in Chapter 11), provide an opportunity for evaluation both by an instructor and his supervisor.

Delegating Control

In dealing with service personnel, laborers, and technicians, standards should be established that can be measured in terms of quantity and quality. The more the employees feel they are participating in the process of control the more successful it will be and the higher will be the morale of the organization. In developing societies personal involvement and concern by the director are often more important because the staff is accustomed to being part of a small enterprise. Its members tend to feel threatened by the impersonal nature of a more structured organization, as development programs expand and the number of employees increases. This problem is usually exacerbated by the absence of adequately trained middle-management and experienced supervisory personnel. Inasmuch as the ultimate purpose of control is to lead to prompt corrective behavior, the participation and support of those directly involved bring quicker and better results.

When the number of employees at the Farm School did not exceed sixty, it was relatively easy for the director and his immediate associates to maintain control through a close personal relationship with all the staff, even though direct supervision was the responsibility of department heads and supervisors. Personal contact provided considerable opportunity for positive motivation and encouragement by top management. This approach is far more effective than one in which the staff members feel that management personnel are continually checking on them. When the number of employees grew to eighty or one hundred, this personal relationship became difficult to maintain.

In any development program the director has the ultimate responsibility, but the more responsibility he can delegate, while ensuring adequate control, the more time he will have to concentrate on program activities. At the Farm School the director was originally expected to sign each voucher. Subsequently this responsibility was delegated to the associate director, and it is currently in the hands of the department heads and the chief accountant.

Implementing Control

A review of the budgeting process at the Farm School indicates the steps that ensure adequate control through appropriate accounting. Just as clearly stated departmental objectives reflect an analysis of the overall plan of the institution, so the budget expresses these objectives in numerical terms. In this way it is the definitive tool for control—the most concrete expression of the POLKA in measurable terms.

The Farm School budget, which covers the period from September 1 to August 31, results from earlier program planning in each department of the school. Every December each department prepares a budget for the year beginning the following September, based on a series of meetings on program planning with the administration. These departmental budgets are combined during January and submitted to the administration in a summary form. Because there is never enough money to meet the needs of each department, the associate director reviews each budget with the department head and revises it in accordance with the available income. In some instances new sources of increased income are sought, and in others program activities are cut back.

Because budgeting must be related to long-term planning, these revised budgets are incorporated in the five-year budget projection, with considerable detail in the first year and less in subsequent years. Once these budgets are completed, they become an integral aspect of departmental management, and each department head is responsible for administering and controlling his own budget. A contingency fund is dispersed at the discretion of the associate director.

Although day-by-day and week-by-week responsibility rests with department heads, the ultimate responsibility is always with the associate director, whose delicate task is to judge when help or interference is required.

Control comes through the monthly financial reports issued by the accounting department. Each report has the following headings: budget for the year, expenses for the month, expenses to date, and balance. These reports are prepared by the tenth of every month for analysis by each department head and the associate director; thus they provide a continuous review, with responsibility at the department head's level. This technique is referred to as "management by exception": the associate director is interested only in reports on exceptions or changes from the budget rather than in minor details that the manager himself must consider.

Every three months the managers submit their estimate of the anticipated year-end results based on the results to date. Thus in September, November, February, and May the budget originally submitted in February is updated and presented to the board, with revised estimates based on a review of both the previous fiscal year and the current year to date. The cash flow statement is equally important because revenues and expenditures are seldom in phase with the budget. On more than one occasion the school has come very close to financial collapse because of its failure to anticipate significant cash flow demands in excess of what appeared in the operating budget.

When authority was transferred from top management to department heads, many staff members questioned the approach, but this approach is appropriate for a number of reasons. As an integral part of their training, delegation ensures that managers feel both responsible and involved. Because they are closest to the problems, they become more cost conscious and provide practical solutions when they feel personally responsible for the departmental budget. By following specific procedures and policies they learn to delegate without losing control.

This system has created a strong sense of team spirit at the Farm School, in contrast to earlier periods when all decisions were made by top management. Although delegation would seem to be a relatively simple process, it requires a capable and well-organized business manager and department heads who are prepared to accept the responsibility and the authority to work as a team. The staff members feel that they have made tremendous progress since this management approach was initiated. Intercommunication and confidence among employees are essential to this process.

CONTROL BY PEASANTS

Control is not as simple for the peasants growing crops or livestock as it is in institutions or among families operating grocery stores or coffee shops. Before World War II, when the shrewd businessmen in Thessaloniki were asked why they did not invest in agricultural operations, they invariably replied that a farm had no storefront that could be locked at night. The function of management training is not only to teach peasants to maintain control so that, like Barba Manoli (see p. 96), they know if any apples are stolen but also to be sure that their apples produce maximum returns at minimum cost.

Production Problems

A group of apple growers at a Farm School course discussed the difficulties of trying to maintain control in an agricultural enterprise. Unlike manufactured goods, farm products such as apples are perishable, which was an even greater problem twenty years ago before cold-storage facilities were used. Furthermore, apples are bulky, hard to transport, and require a relatively large storage space. A good apple harvest creates a glut on the market, and prices drop below production costs because the farmer must dispose of them. In a poor harvest, prices are high, but few farmers produce enough to make a significant profit. For every new pesticide marketed to protect the apples a resistant strain of pests seems to develop. As labor demands are seasonal it is virtually impossible to find workers when they are needed. They concluded by saying that they are all exploited by the merchant, who makes more profit on their crop in a few days than they do working for a whole season. Development workers should help peasants organize production and processing techniques so that they can maintain better control in the marketing of their products.

Barba Manoli, a peasant in a village near the Farm School, maintained sufficient control of the apples stored in his warehouse to realize that young children were stealing a few each day by sticking them with a sharp pole between the bars. He put up a sign stating that the apples were covered with poisonous pesticides. The next day he found a note from the children who stole his apples. "Don't worry, Barba Manoli. We wash the apples before we eat them."

Successful farmers understand that the "eye of the master" is the best fertilizer for his farm. Many farmers, even today, visit their fields only when they plow, plant, fertilize, or harvest. Greek farmers, whose farms are made up of an average of eight acres subdivided into nine or ten plots in different locations, find it difficult to check their fields regularly. In recent years the government has undertaken a voluntary program of land redistribution in many villages, which has integrated the villagers' holdings into one or two plots. The redistribution makes it easier for the farmer to maintain control over the plots and results in more efficient farming.

A marketing plan that includes quality control is another mark of a professional farmer. Only recently have fruitgrowers begun to overcome the widespread tradition of packing inferior fruit underneath those of high quality. For years the producers were at the mercy of merchants who encouraged this practice, undermining the reputation of Greek products in central Europe, where Greece exports most of its fruit. Peasants have recently formed cooperative groups to include marketing in their production process, and with the help of the government and the national cooperative movement they have been able to improve their image.

Only a handful of competent managers among the peasants in each village calculate the cost of producing their crop and control production expenditures. Many have a general idea of costs and assert that they keep track of their expenses, but they have little grasp of hidden costs such as depreciation and inflation. It is important to teach farmers not only how to keep records but also how to interpret them because proper use of records leads to adequate control.

Greek villagers have traditionally been victims of high interest rates. For years they were taken advantage of by the grocer, the tobacco merchant, and the itinerant salesman, who would either lend them money at exorbitant interest rates or sell them goods on credit at prices over which they had no control. When they have no ready cash, peasants often purchase goods on credit, which the merchant registers in a book. As the merchant is the only one who keeps a record, the villagers have no way to knowing what the credit is costing them, particularly when more than one member of a family shops.

Controlling Production

Many farmers have had to learn to use family labor effectively because workers have become increasingly scarce. The Farm School surveyed village dairy, poultry, and greenhouse operations to determine the optimal unit that could be independently operated by one man and his family.

This survey was prompted by the shrinking rural population and the increased cost of casual labor.

The school's twenty-four-cow dairy unit, which can be operated by a man and his wife very easily, was built as an outgrowth of this study. A recently completed mist-propagation greenhouse is another example of a single family enterprise that does not require outside labor. A sixteen-thousand-hen poultry unit was also designed on the assumption that it would be managed by one man and his family, but because it requires technical skills for operating and maintaining the automated equipment, the unit is not really an appropriate project for a village family.

Few peasants have learned to distinguish between capital and income. A peasant who attended several short courses at the Farm School sold plots of land over twenty years to invest the money in a new enterprise. As he did not keep records of his expenses, he sold his products below the costs of raw materials, labor, and overhead. Only three or four years later, when he had no more fields to sell, did he realize that he had consumed his working capital. The nomadic shepherd clans of Greece, who were better businessmen, always maintained an unwritten law that the dowry given when a daughter marries should never come from the sale of sheep, which they considered capital; the dowry was accumulated in gold coins from the sale of milk, cheese, wool, and lambs. Only when shepherds decided on a more settled life-style did they sell their sheep and use the money to buy land, which they regarded as another form of capital.

A plan of activities around which agricultural operations were organized proved helpful to many successful farmers. In Greece this schedule was not kept by calendar dates but was related to traditional saints' days. Every master farmer knew that certain jobs must be completed by a specific time, such as the Day of the Virgin (August 15), St. Demetrios' Day (October 26), or St. Anthony's Day (January 18). The difference between the nikokiris and the kakomiris was that the former knew the deadlines and kept them whereas the latter remembered them but very often let them slip past without taking action.

Well-organized farmers profit from records of their past experience to help them plan and control their future program. When villages were remote and life less complex, boys accompanied their fathers to the fields and absorbed a deep understanding of seasonal activities. But a variety of distractions and the need to go away to high school have kept young people from day-to-day farm operations. Those who do return to farming must work out a calendar of events that can guide them in their planning and help them to control their agricultural operations.

NEW SYSTEMS OF CONTROL

A variety of systems are now available to help the farmer maintain control in operations. In a very comprehensive system being developed by the Swiss Extension Service, loose-leaf reference folders are coded by types of cultivations or enterprises and by the activities related to each. Thus, the field crops are broken down into winter cereals, spring cereals, sugarbeets, potatoes, corn, colza, feverole, forage beets, and forage production. Each crop is then analyzed and color coded to provide information about culture, varieties, pesticides, harvesting, and storage.

For each crop the service provides the farmer and the advisor with a detailed guide listing production activities required, dates, the stages of the plant's development, factors to beware of, and alternative solutions to problems that might arise. If a farmer's winter crop is destroyed by frost, this guide helps in such decisions as whether another crop should be planted and what crop can or cannot be planted depending on the herbicides and pesticides used. It also tells the farmer what stage should have been reached in the activities, which helps the farmer evaluate progress relative to a norm. The guide provides him with continual control from crop selection until marketing.

Similar services are provided in Illinois by the Farm Business and Farm Management (FBFM) systems. At Michigan State University, this process has been computerized with its telefarm approach. The staff members from the Department of Agricultural Economics at Oklahoma State University have been developing a similar program in Honduras by collecting information on specific farms and attempting to design a general model based on this material.

Many economists doubt that the peasants can maintain sufficient control of their enterprises to become master farmers. The peasants' success will depend on the development workers' ability to instruct them in the use of the variety of systems now available for effective control and on their willingness to use these systems properly.

Hodja decided to teach his donkey to eat less during a year of drought. Each day he reduced the amount of feed until one morning he found the donkey dead. When Hodja started lamenting, his neighbor asked him what was the matter. "I had just taught my donkey to get along without any food," said Hodja, "and he died."

9

Adjusting for Flexibility

- When should the process of adjusting be implemented and how?
- What relationship does it have to the rest of management?
- Can the mechanism of adjusting be taught to a peasant?

Adjustments are necessary in any approach to ensure flexibility in the implementation of the POLKA. Adjusting is the process by which corrective action is taken, based on observations when controls are applied, to ensure that the original objectives are attained or that they are modified to changing conditions. Hodja's idea of reducing the amount of food consumed by his donkey reflected intelligent management, but he failed to make a corrective adjustment.

Two unsuccessful attempts to introduce beef cattle into Greece illustrate the importance of maintaining flexibility in development programs. For many years, visiting experts had urged the Farm School to try out beef cattle in Greece. The opportunity arose when President Dwight D. Eisenhower presented an Angus bull and two heifers to the school to be exhibited at the Thessaloniki Trade Fair, where they were viewed by thousands of people.

When the Fair ended, a feeding trial was organized at the school using the offspring of the original stock and a few crossbred animals. Four years of trials with the Angus proved that no market existed for high-priced marbled beef and that the pasture land for this kind of cattle was insufficient in Greece. The animals were slaughtered to eliminate further losses and to discourage other farmers from making the same mistake.

In another venture a company formed by Greek and Texan farmers imported five hundred head of beef cattle. The animals grazed on mountain pastures that had been used by sheep for centuries. Barns and other facilities were constructed with loans from the Agricultural Bank. During the first summer wolves attacked the calves, pastures dried up, and disease struck some of the herd. But the company's biggest problem was finding markets for the high-quality beef because cheaper meat was available from Argentina and Australia. The company's debts grew to more than a million dollars, and it was forced to declare bankrupty. By the time the managers realized their problem the commitment in livestock, buildings, and shipping costs was too large to make an adjustment. As compared to the first feeding trial the larger unit was too complex to be terminated without serious economic or social loss.

Adequate adjustment is only possible when a project has been characterized by proper planning and organizing, able leadership, and effective control. Institutional managers often fail to recognize the dynamic nature of management: They think of it as rigid rather than as a changing, active process. Introducing flexibility to a development program and teaching farmers to adjust are two of the most difficult tasks of the development workers. Such workers should be continually alert to every aspect of the operations they are managing and respond according to the needs.

ADJUSTING IN DEVELOPMENT PROGRAMS

A view of the complex nature of management that has been helpful to the administration of the Farm School is illustrated by the following statement.

> Every successful institution or organization needs two kinds of people—an administrator and an innovator. The administrator organizes the operation and ensures that it runs smoothly, while the innovator—the ideas man—constantly seeks new ways to resolve problems. Without the administrator the organization soon disintegrates and rapidly heads for bankruptcy. Without the innovator it will keep repeating the same programs until it no longer serves a purpose. Unfortunately these two types are directly opposed. The administrator will protect his systems and forms unyieldingly, while the innovator regards them as a challenge to make changes. To the administrator the innovator seems like a dangerous lunatic, while the innovator finds the administrator stuffy, unprogressive and obstructive. The measure of an effective organizational leader is his ability to maintain a balance between the two, protecting the impulsive, creative innovator from the just wrath of the conservative administrator.[1]

The manager cannot be satisfied just by establishing an organization to promote development: He must make sure that it is in constant tension, maintaining a balance between the administrators and the innovators.

Although private institutions are usually more adaptable than public organizations, they still suffer from inflexibility—the product of institutionalization and self-interest among the staff. One factor that contributes to this rigidity is the buildings themselves. As activities increase and buildings are added, staff attention is diverted from training programs to operation and maintenance of the physical plant. One YMCA secretary who dreamed of inspiring idealistic young men and women through his organizational efforts eventually resigned because he found he was spending most of his time looking after buildings rather than serving youth.

The Farm School has also been burdened by too large a plant. Dr. John House started the Farm School with one building, and now it has fifty, many of which would be difficult to remodel and currently require constant maintenance. Just as the physical factors lead to inflexibility so staff using these facilities fall into daily routines that are difficult to modify.

In one developing country the Ministry of Agriculture and a foreign aid group designed a short-course center attached to an agricultural school. The project was expected to help the school reach out into the community and to bring the community to the school. The staff decided to hold the first short course even before the facility was completed. It was an overwhelming success with lectures, demonstrations, field trials, and visits to the student projects. At the end of the course the trainees, who were accustomed to spending most of their time out of doors in the villages, were asked if they had any suggestions. They replied that they would prefer to listen to the lectures sitting on the grass than inside the classroom. Having felt uncomfortable inside the center, they were suggesting an adjustment even before it was completed.

Equilibrium in Management

The Board of Trustees of an organization approves policy recommended by the administration, the staff of development workers implement the resulting programs, and the financial officer prepares and administers the budget. It is the responsibility of the chief administrative officer to maintain a balance among the three. If any one of these elements dominates the other two, some type of adjustment will be required.

One approach to the problem of maintaining management balance, devised by Professor Tad Hungate, the comptroller of Teachers College at Columbia University, has been very helpful to the Farm School administrators for many years.[2] It is also particularly useful in private overseas development work in which problems of communication often develop among the Board of Trustees, development workers, and the finance director.

According to Dr. Hungate's theory, the three elements of institutional management should be in a state of dynamic equilibrium. The director of the program, the chief administrative officer, is responsible for coordinating the three groups involved in management so that each understands the other's plans and objectives. At various times one of the three elements may become dominant over the others, causing an inbalance in the organization. When board members attempt to dominate programs, staff activities, and budget details, employees become demoralized and friction grows within the institution. In situations in which the staff becomes dominant, it is inclined to be guided by the needs of the program as it sees them, without concern for either policy considerations or budgetary limitations. Trustees become frustrated, lose interest, and fall short in their efforts to raise the funds needed to finance the program. By disregarding the budgetary implications of their actions the staff brings the organization to the verge of bankruptcy, causing outsiders to lose confidence in it.

When the fiscal officers insist that financial considerations must dominate both policy and program, the staff and trustees, none of whom feel they play any part in the decisionmaking process, become demoralized. Financial officers in such positions often take pride in creating large financial reserves but fail to understand the damage they are causing to the institution.

One challenge for the director of a training institution is to continually evaluate the relationship among these three forces and adjust any imbalance to ensure a harmonious relationship. Anyone who has worked with such programs can cite several that have failed in their mission because the director has not been able to fulfill this role.

Adjustments in Staffing

A great mistake institutions make is to assume that the abilities of some staff members will remain static, whereas they usually continue to expand. Dr. Francis C. Byrnes of the International Agricultural Development Service distinguished between a worker's frustration when his abilities are not adequate for his job and the apathy of the more capable worker who is not stimulated by a job that demands far less than he can contribute.

With the passage of time, the abilities, interests, and amibitions of young staff members grow beyond the requirements of their position. The employer then has two options: developing new programs that will keep challenging the employee, or suggesting that the employee move to another institution that can better use the broadened interests and ambitions.

At one time the Farm School staff tended to be insular and resented the members who resigned to join other institutions. It eventually recognized that there was not sufficient scope for all the capable young staff members and that some should leave and others be promoted. The morale of those who remained at the school improved once they accepted the idea that a moderate staff turnover was normal and beneficial to an institution.

TEACHING FARMERS TO ADJUST

Two obstacles make it difficult for farmers to make rational adjustments to fit changing circumstances. First, they have a characteristic inflexibility that grows out of traditional conservatism and distrust of innovation. Second, emotionally immature farmers tend to be easily swayed by circumstances when dealing with problems. The common phrases used by the kakomiris ("It can't be done," "You know who I am," and "I understand, I understand" when they do not) are indicative of their inflexibility. The challenge of the development worker is to help them understand the concept of management by objectives and self-control, so that they can learn to establish clear goals while maintaining the flexibility needed to adjust to changing circumstances.

Practical Experience

Peasants learn the significance of flexibility in management from personal experience as well as through training. The master farmer is closely attuned to the slightest changes in crops and animals and is able to respond to their needs. The villagers must be taught this flexibility through practical experience in farming, following the Confucian precept, "I do, I understand." A group of Farm School students raising feeder pigs and sows discovered that the sows were growing fatter and the baby pigs thinner. Their concern, originally discussed by the students and their advisor, spread to their class and the whole school. After some time they realized that they were feeding the sows' ration to the baby pigs and the feeder ration to the sows. By the time they made the adjustment, they had consumed their potential profit but were much richer in experience.

The Israelis have found that the best way to ensure continual adjustment is through field trials. Anyone visiting Israel to study agriculture will be shown field trials carried out by farmers, agriculturists, research stations, the university, or a combination of these working together. The Israeli approach to developing new cultivation projects bears study by other countries. When the Israeli Ministry of Agriculture took the decision to market cut flowers for export on a large scale, a group of top agriculturists was sent to the United States and several European countries. The ministry insisted that these agriculturists should work while in the foreign countries, besides studying how the industry was operated, to acquire the necessary practical experience.

Simultaneously, the ministry organized a program in Israel to train scientists who specialized in diseases, marketing, and other phases of cut-flower production. This group cultivated experimental plots and greenhouses, which became practical training grounds for agriculturists and farmers who had not been abroad. In both the kibbutz and the *moshav* (the cooperative villages), through the joint efforts of specialists, agriculturists, and farmers, the program was continually reviewed and adjusted, leading to improved techniques. In a very few years cut flowers have become an important export crop.

A somewhat similar approach has been taken by the Coffee Board in Colombia, which has experimented with a variety of agricultural crops in addition to coffee, to find ways of making the farmers less dependent on this one cash crop.

Common Market Programs

Two Common Market regulations help the European farmer to adjust to changing conditions. Regulation 159 provides for a detailed study of each farm unit by the farmer and the agriculturist in the area. Once this specific plan has been drawn up, funds are granted for long-term loans with interest subsidies that can be used for land improvement, investment in fixed assets, and the purchase of livestock. A subsidy is also granted for keeping accounts, which helps the farmers maintain adequate control and make adjustments in their operations.

Common Market Regulation 160 provides for socioeconomic advisors, whose job is to advise farm families. They also help less productive small farmers decide whether they should make a radical adjustment and either retire from farming or obtain training in a nonfarming occupation. This service is particularly important for farmers over fifty-five years of age. In this way land being marginally farmed becomes available to master farmers, and marginal farmers are encouraged to change their mode of living. This is one of the most controversial aspects of the farm modernization scheme under the Common Market. People question the right of outsiders to advise any farmer to relinquish the joy and stimulation of farming for less satisfying work elsewhere.

The Adjustment Process

Helping farmers to understand the process of adjustment has been a concern of the Farm School staff for some time. Some of their guidelines for farmers beginning a new activity, brought out at a 1979 staff meeting, might be helpful to development workers: "New programs should begin in a small way. It is much easier to make adjustments in a pilot project than in commercial operations. Too many projects are started on a grandiose scale rather than on a limited experimental basis which allows for flexibility. The simpler a program the greater will be its chances of success."

Development workers in foreign countries should focus on helping peasants modernize within their own culture rather than concentrating on foreign technology. J. Wellington Koo, the Chinese ambassador to Washington after World War II, told a group interested in development work that what China needed was neither westernization nor Christianization, but modernization.

Introducing foreign equipment without adapting it to a new cultural situation creates unimaginable problems. The local people often reject it out of hand. Spare parts are seldom available, and it is difficult to find villagers with the technical expertise for maintenance. Although there is great value in studying the accomplishments of other countries, foreign technology must be adapted to local situations with caution.

Development should be thought of as an evolutionary process growing out of a series of adjustments to changing circumstances. Farmers who are encouraged to try innovative methods must understand that their production must be integrated with sources of supply, labor potential, and marketing possibilities. They should be encouraged to keep returning to the basic steps. Master farmers in Ireland teaching apprentices management discovered that they benefited greatly from analyzing their own methods in order to explain them to the apprentices. They reported that a constant review of their accomplishments and their methods helped them recognize the importance of the basics in the growth of their operations.

As farmers make progress in new enterprises and acquire more self-assurance, they tend to become overconfident. They grow lax in their planning and organization and in their systems of control—which often leads to disaster. Three years after the Farm School began an experimental program in broiler production with its graduates, more than half had failed in the business because they had become careless after their early success and had failed to prepare for setbacks such as disease, price fluctuations, and marketing difficulties. The duality of a person's fortunes was acknowledged by the ancient Greeks, who cited two inseparable lesser deities of Mount Olympus. The first, Tichi (fortune), gave happiness, whereas the other, Nemesis, brought divine retribution. The ancient Greeks believed that whenever one was present the other was not far away. Peasants learning to adjust must adapt to the fruition of both their most optimistic expectations and their greatest fears.

The flexibility that ensures adequate adjustment in management is seldom present in organizations or among the peasants whom they train. Institutions become weighed down by buildings, administrative procedures, and the self-interest of the employees, who become fixed in their ways. Peasants are hesitant to accept change for fear of losing what little they have. Thus one function of management training in developing countries is to help both development workers and farm families learn to adapt to changing circumstances in order to keep up with progress, both in their own society and other cultures that directly influence their own. For this reason the Farm School emphasizes problem solving because adjusting is an ongoing process of adapting a set of goals as well as the means of attaining them.

PART THREE

Heads, Hands, and Hearts

The classic approach to education in developing societies emphasizes factual knowledge. A well-balanced training program, however, should equip the progressive peasant with adequate agricultural knowledge, skilled, calloused hands able to undertake a variety of tasks, and an attitude open to outside suggestions. The Greek concept of an open-hearted person is one who relates well to others and is receptive even to unfamiliar ideas. Although training centers teach management skills and problem solving, essential attributes for peasants striving to become master farmers, they also must build self-esteem—the pride a peasant takes in knowing who he is and what he wants from life.

The importance of these elements in training master farmers and observations on the best methods by which to teach them are examined in the following chapters. They are based on the perceptions of individuals in various cultures and institutional settings who have devoted many years to helping peasants cultivate their heads, their hands, and their hearts, as well as their gardens and fields.

One day the village teacher told Hodja that he had decided to travel across the land to seek additional knowledge. When the young man asked him what kind of people he should look for, Hodja recalled some wise words he had once heard in the bazaar:

> He who knows not and knows not that he knows not is a fool. Shun him.
>
> He who knows not and knows that he knows not is a child. Teach him.
>
> He who knows and knows not that he knows is asleep. Awaken him.
>
> He who knows and knows that he knows is wise. Follow him.

Hodja paused for a moment and then continued, "But you know how difficult it is, my son, to be sure that the one who knows and knows that he knows really knows."

10

The Role of Knowledge

- How much technical knowledge does a master farmer require?
- Who should decide what the master farmer needs to learn?
- What is meant by teaching knowledge?

Dr. Francis C. Byrnes introduced the Farm School staff to the concept of the squares of knowledge, which is not unlike Hodja's evaluation of individuals. The four squares correspond roughly to four categories of people: those who know and know that they know, who are wise; those who know and do not know that they know, who are humble; those who do not know and know that they do not know, who understand their limitations. "But the people you have to beware of," said Dr. Byrnes, "are those who do not know and do not know that they do not know."

When a Greek entomologist who had just returned from the United States was asked what he had learned during the six months of his stay, he replied, "Three words—*I don't know.*" He reported that his colleagues in the Ministry of Agriculture were amazed when he said "I don't know" in answer to a question and that he would find out. "Now when I answer their questions," he said, "they know that I know." University-trained agriculturists feel it would be embarrassing to admit that they do not have all the answers. It does not seem logical to expect peasants to admit ignorance when well-educated people refuse to concede that they do not know. An important aspect of training in developing countries is to encourage everyone involved in a program to feel free to acknowledge ignorance of any particular subject.

DEVELOPING TEACHING TOOLS

In Greece today there are shortages of audiovisual aids and other tools to assist teachers; this problem is even more crucial in Third World countries. Only now after eighty years is the Farm School placing major emphasis on designing teaching packages for agricultural schools and short-course training centers. The Greek experience would indicate that one of the top priorities for technical assistance programs in Africa, Asia, or South America should be regional centers for designing teaching packages.

Teaching Packages

At Michigan State University the staff members of the Soils Department have prepared a series of lectures on cassettes accompanied by slide carousels. A room similar to a language laboratory is divided into booths containing small tape recorders and projectors. As part of the course in soils, the student reads the script, hears the voice on the cassette, and studies the slides. From time to time the cassette tells him to go to the laboratory next door to perform specific tasks. When he completes these tasks, he returns to his machine and resumes work. This method

	KNOW	DON'T KNOW
KNOW	1	2
DON'T KNOW	3	4

enables the student to study at his own pace, reinforces his learning with visual and practical aids, and reduces teacher time.

Dr. Fritz Kramer at CIAT in Cali, Colombia, designed a series of similar teaching units each costing approximately $50. They are an invaluable tool for both teachers and students in an agricultural development program. Each unit, which can be used independent of an institutional instruction program, consists of one or two carousels of slides, cassettes, a book of instructions, a text, and a series of questions. Although it takes two to three years to prepare such units, they are well worth the effort.

The state of Oklahoma operates a Curriculum Materials Development Center at Stillwater, Oklahoma, which is an educator's dream. The Center prepares materials for agricultural education and other vocational programs in Oklahoma. The staff has designed a series of teaching packages that incorporate the very best principles of learning. Short courses are held annually, under the direction of highly qualified professionals, to teach others how to prepare teaching packages. Similar centers are operating in other states, including Ohio, Arizona, Texas, and Georgia.

The University of California at Davis operates a Curriculum Materials Library in the Department of Agricultural Education that collects teaching packages and teaching units prepared by other states and by private industry so that students at the school and others may learn what is being done elsewhere. A similar center is being established at the Farm School.

John Deere, the agricultural machinery company, has developed an excellent series of materials for teaching farm machinery maintenance and repair. Their Fundamentals of Service (FOS) Manuals carefully break down each operation that a farm machine repair technicians must learn. Because farm machinery is independent of cultural deviations, the accompanying slides, transparencies, and other materials can be used in other countries without having to substitute new pictures. The only requirement is that the texts and the picture subtitles be translated into the user's language.

Subscription Services

People working in developing countries have difficulty obtaining the information they need in all sorts of disciplines. Textbooks, periodicals, and other reference sources are not readily available in developing countries, where there are few libraries. The late Dr. Fernando Monge, director of the library at CIAT in Colombia, defined a library's function as "to put information in the hands of the user at the time and place when he needs it." Dr. Monge, who was a communications specialist and not a librarian, was distressed by the institutionalization of libraries, which serve only those who visit them, and evolved a practical and imaginative approach to circulating information to most countries in South America.

The CIAT library subscribes to all scientific periodicals relating to CIAT's four areas of production—cassava, beans, pastures, and rice. Once a month the library makes photocopies of the table of contents of each of these and sends them to subscribers. A token subscription is charged to ensure that people really want the service. The subscriber circles titles that are of interest and returns it to the library. The articles are photocopied and sent to the subscriber for a minimal fee. The Farm School is currently implementing a similar system to make the contents of selected agricultural periodicals available to graduates and other friends throughout Greece.

EVALUATING TOOLS

A member of the Farm School faculty who visited an agricultural school in Tanzania watched thirty students sitting at their desks writing notes dictated by a teacher because the school had no textbooks. What motivated the students to memorize the material was their desire to pass examinations for government positions. Fortunately a comprehensive program of practical instruction partly made up for some of the wasted classroom time. The visitor was convinced, however, that even if they did obtain the appointments they would require a great deal of in-service training with farmers to compensate for the deficiencies in their theoretical training.

The same visitor was impressed by an agricultural school in another part of Tanzania that had better farm equipment than many countries in Europe. Foreign agricultural missions from Asia, the United States, and European countries competed with each other in their assistance programs to this school. Although the availability of buildings, machinery, and livestock contributes to the training of agriculture students, it does not automatically solve agricultural education problems. Trainees need to acquire fundamental knowledge, develop new skills, and understand the elements of management in order to become more effective farmers. Professional educators are more apt to think in terms of hardware such as projectors and audiovisual equipment than the software for teaching packages, slide lectures, and film strips because these must be prepared locally. The most important aspect of the training program is to design experiences for the individual trainees through which they will acquire the necessary knowledge.

On more than one occasion the Farm School has outstripped the peasants' capacity for assimilation. Immediately after World War II a model of a one-hundred-cow New England dairy barn was built at the school with government grants. It is an example of the architectural imperialism of which expatriates in development are often guilty. Recently the school built a twenty-four-cow model unit of the simplest construction. When villagers saw the small barn, they asked for the plan so that they could copy it; no one has ever been interested in copying the larger building.

An even simpler six-cow dairy unit, which the students manage themselves, was built for their training. As the school pays the students for the milk and charges them for the feed costs, they feel that this operation is their own. The students also use a smaller milking unit consisting of a rubber udder on a frame, which is an exact replica of a cow's udder and can be milked either by hand or by machine. By practicing on a rubber udder the student can learn without damaging the cow or being kicked.

DESIGNING TEACHING PACKAGES

Development workers constantly ask how they can train farmers more effectively. Although the Farm School has not been completely successful in this area, several new approaches have been helpful in improving the instruction. The school is implementing a Curriculum Materials Development Center, which has already produced two teaching packages—in dairying and in farm machinery—and other teaching packages and materials from abroad are being assembled. Funds are being sought for a project to adapt audiovisual material to local conditions and translate scripts into Greek. It might be useful to review how this program was accomplished.

The first step in developing teaching packages was to hire a communications expert for a three-year period who was experienced both in teaching vocational agriculture and in designing teaching packages. Next the school selected Greek associates who could work with the consultant and continue after he had left, which would assure that the work would be thoroughly Greek in its approach.

Probably the most difficult part of the project was to agree on a standard form for the teaching package. Faculty members took a long time to understand how helpful the teaching package would be for both teacher and student. In the soils course at Michigan State the teaching package almost replaced the teacher. The CIAT course in Colombia was a self-teaching unit for adults. In secondary schools, however, the teaching package is a tool, not a substitute for the teacher.

The model teaching package, developed at the Farm School, which is not unlike the Oklahoma model, is based on the following outline.

The *objective* states what the trainee is expected to learn. The emphasis is on performance-based objectives, which describe what the student will be able to do when he completes the course. This approach is particularly important in developing countries, where teachers tend to concentrate on what they want to teach rather than what the student is expected to learn. It is hard for them to understand that if the student has not learned then the teacher has not taught.

Bibliography and instruction materials list the equipment, background references, and audiovisual aids needed by the teacher.

Rationale explains why the lesson is important to the trainee, to gain his interest and involvement.

The *suggested teaching activities* outlines the method of teaching and the steps that must be followed to ensure that the student acquires the desired knowledge, skills, or attitudes.

Participants' problems states some of the concerns and questions related to the subject, as a means of encouraging the student to discuss the problems as he sees them.

Questions to be asked by the teacher are listed to assist in assessing the student's understanding of the lesson.

Application problems provide practical exercises for the participant to reinforce learning.

One day the Hodja came to the mosque unprepared for his sermon. He asked the audience if they understood what he was going to tell them. When they all replied no, he told them if they did not understand then there was no point in telling them, and he sat down. The following week when he asked the same question they all replied yes. Since they knew what he was going to say, he told them there was no point in saying it, and again he sat down. By the third week half the audience said yes and the other half no, to which Hodja bowed politely and asked those who knew to tell those who did not.

Self-evaluation consists of comprehension quizzes in multiple-choice or true-false questions. The answers are provided to help the student in self-evaluation.

The *summary* is a review of the lesson that can be given either by the teacher or by a student.

The teaching package should be used as a guide, not as an outline. Teachers should be encouraged to supplement the materials with their own personal examples and questions to improve the relevancy and the effectiveness of the instruction. The final and essential step in every lesson is to determine whether the participants have actually achieved the objectives. Instructions for teachers on the use of teaching packages are included in Appendix B.

SUPERVISING AND TRAINING INSTRUCTORS

Different countries and institutions have varying philosophies for training and supervising instructors. Some U.S. states give a great deal of freedom to their teachers to develop programs, whereas others require them to follow a detailed curriculum. In Greece the material is defined in great detail by the Ministry of Education, so that every teacher will cover the same prescribed body of information within the given time frame.

These two methods of educational supervision are not unlike the X and Y approach of Douglas McGregor, an early theorist of business management.[1] The type X manager is authoritarian and has little faith in human self-motivation. The Y manager tends to be democratic in administration and allows self-expression among the workers.

In Greece the Ministry of Education's high school programs tend to be X-oriented, with a rigid curriculum expected of all teachers, whereas the Ministry of Agriculture gives greater freedom for the students and the instructor in its short courses, based on the students' needs. Just as few managers are entirely X- or Y-oriented, so most educational supervision falls somewhere between the two extremes.

One of the most important teaching techniques is using "those who know to tell those who do not know," as Hodja said. A center in Ireland has introduced the concept that instructors in short courses should think of themselves as promoters of development rather than as teachers.[2] The most effective teaching comes from the best students. The Dutch feel that the best teachers are the farmers themselves.[3] As one teacher in Holland remarked, "A group of trainees each with his own background is bound to have a combined knowledge greater than that of the instructor."

When the Hodja came to the mosque one day he noticed an unknown goatherd below the speaker's podium. The man had walked for three hours to attend the service, leaving his goats up in the mountains. While Hodja was preaching he saw how deeply the man was moved. As the sermon progressed the goatherd pulled out a rag and started wiping the tears that were flowing down his face. Hodja became more and more excited. After a final burst of enthusiasm he sat down. At the end of the service he asked the frightened goatherd what had moved him so deeply about the sermon. "Oh Hodja, wise man," replied the man, "last week my best billy-goat died. The more I watched you talk with your beard dangling the more I thought of my billy-goat and the sadder I became."

It is often easier to train mechanics to be teachers than to train teachers to teach mechanics. This was discovered many years ago in setting up farm machinery training courses when very few expert technicians were available. The university agriculturists were well-trained in theoretical aspects of machinery, but they had no practical experience. In the end the best trainees from a number of courses were chosen, trained to teach, and hired as instructors. They spoke the same language as the farmers, understood their problems better than the agriculturists, and were much closer to them.

The Ministry of Agriculture had a comparable experience in teaching land redistribution to the farmers. They organized a three-day short course in which the first day was devoted to discussing the subject theortically. Most of the farmers were not only not interested but were hostile to the idea. On the second day they visited a village in which land redistribution had already taken place. The villagers were able to provide first-hand information about how the system worked from their personal experience, and these contacts convinced the trainees of its benefits. By the third day the trainees had a new attitude and were eager to acquire additional information to organize land redistribution in their own villages.

MOTIVATING THE TRAINEES

Motivating the students to learn is an essential element of the training process, particularly when working with adults whose minds are on various personal problems, as illustrated by Hodja's experience with the goatherd. Only as teachers mature do they understand that even the students who seem attentive are often daydreaming rather than absorbing what is being said. Their challenge then is to find ways to get past "Hodja's beard."

Short-course teachers often fail to capture the students' attention because they mistakenly take the trainees' interest for granted. Some of the best short courses at the Farm School over the years have been in recreation. On the first day of the one-week courses trainees are told that they will be required to put on a recreation program by themselves for the students and staff of the Farm School. From early morning until late at night they work like beavers developing materials and innovations, turning to the instructor for additional ideas. Their course is an active beehive at work because they know from the beginning that they themselves will be the performers.

Many short-course teachers determine the material that they teach by their own interests rather than the needs of the students. At Michigan State and at Oenkerk in Holland questionnaires, sent out in advance of short courses, list the possible subjects from which the student may choose. Trainees are asked both to state their needs and to evaluate their own competencies in each of the proposed subjects.

Few teachers are able to fully stimulate trainees' thinking. Projects operated by students for their own profit have been particularly helpful teaching tools for capturing students' interest. When a whole class operates a dairy or pig unit, interest is generated but staff leadership is required; in contrast, student projects managed by a small group stimulate the students to frequently seek help from their teachers. They even get up before dawn to work on such projects. Requiring students to work in the school's large production-demonstration units as laborers is the least successful approach to practical training and requires constant supervision by the staff yet has only marginal results.

Teachers often fail to incorporate adequate active participation of the trainees in their courses. One-week pork preservation courses for village women were among the school's most successful programs. On Monday morning the women slaughtered a pig, and by Saturday afternoon they carried home with them a can of pork, pieces of ham and bacon, a tin of lard, and some soap, all of which they had prepared themselves at the school. Invariably the women were challenged to process the whole pig in a week.

The Oenkerk School in Holland provides a series of one-week programs based on the activities of the students. It is assumed that they will learn through what they are doing. The emphasis is entirely on participation in production enterprises rather than on listening to instructors. The trainees feed the livestock, milk the cows, work in the pasteurization plant, bottle the milk, help in the production of different varieties of cheese, and assist in the packaging and marketing so that they have the practical experience of every step of the dairy process. A similar approach is used in teaching technicians to pedicure dairy cows. The trainees spend their training time beginning with hooves from dead cows and reaching a point where they practice on live animals in the school's dairy unit.

Reinforcing appropriate behavior by the trainees is vital: They must feel rewarded for discovering correct answers. Such a positive approach to training is far more effective than a negative one in which students are motivated by the fear of punishment. Unfortunately there is a tendency in developing societies to admonish and humiliate trainees rather than to encourage them through positive support. The Charles Dickens approach of rapping a child's knuckles with a ruler or standing him in the corner for giving wrong answers is as harmful for adults is it is for young children.

A philosophy for teaching knowledge might best be summarized by describing two approaches to teaching students about the kiwi, a fruit unknown to most Greek farmers. In the classical approach the instructor would begin by describing the kiwi and drawing it on the blackboard or showing a picture of it. The students would then be expected to study the cultivation practices in a book and write a report or pass an examination.

In contrast to this procedure the master teacher would encourage the students to grow kiwi plants themselves, keeping notes, making drawings, and taking pictures at various stages. Eventually the students might grow enough kiwis to sell them and make a small profit. By this time they would not only know a great deal about the kiwi fruit but also be able to teach others. Even in a poor country like Tanzania all the tools required to follow these steps are available except possibly the camera. But tools are not the most important part of the teaching process: The challenge is to motivate students to learn through their own activities, giving them the satisfaction of feeling that they have taught themselves.

To teach properly the teacher needs teaching packages that use a variety of methods, including activity by the student and sharing the experiences of other students. Acquiring knowledge is clearly an important element in the training of the master farmer, but in most development programs it has been the most overemphasized aspect of the process. It is far more significant for the trainee to understand his limitations, know where he can obtain additional information when he needs it, and use his knowledge productively in dealing with the variety of problems facing him. These abilities distinguish the master farmer from the kakomiris.

11

Teaching Competencies

- Why is it necessary to teach competencies?
- How should they be taught?
- What is the basis for deciding which skills to teach?

In Hodja's village there was a Greek community with a priest who enjoyed playing chess but had no one with whom to play. One day the priest decided to teach Hodja the rudiments of chess, following which they started a game. Before they began he crossed himself and checkmated the Hodja after a few moves. The next time they played the priest crossed himself and again won. After several games in which Hodja always lost, he turned to the priest and asked him whether if he crossed himself before each game he too might win. "Yes," replied the priest, "but first you have to learn to play chess."

Teaching competencies ensures that peasants acquire the skills to deal with their problems rather than wait passively for prayers or outside experts to solve them. No single maxim is more useful to those working in development than one attributed to Confucius:

I hear, I forget.
I see, I remember.
I do, I understand.

This was the central theme in a training program for short-course directors, given at the Farm School and sponsored by the Ministry of Agriculture, and was translated and printed for use in centers throughout Greece. Whenever the Farm School faculty discusses its educational program, this formula becomes the criterion for evaluating effective teaching. Despite this emphasis on doing, the casual visitor is still apt to see teachers talking while students listen or demonstrating while they watch. Although an observer can easily point this out to teachers, they find it extremely difficult to change their traditional lecture approach.

During the late 1960s when the school was seeking accreditation from the Ministry of Education, the faculty tended to place too much emphasis on subject matter. This approach contrasts to that used at the California State Polytechnic University at San Luis Obispo, which many years earlier had developed a core curriculum of one hundred and fifty basic skills that each student was required to learn before graduation. The *National Agricultural Competency Study* provides an analysis of competencies in agriculture[1] and should be a basic reference for every training center and agricultural school throughout the world. The Farm School has prepared a list of three hundred and seventy-five basic competencies that students should learn. Although each student does not acquire all the skills, it is important to clarify which ones should be included as a part of any specialization. A sample of these competencies indicates the variety of skills required in the various departments of the school.

Students will be able to:
- graft a small tree
- identify the parts of a tractor engine and explain their purposes
- demonstrate the proper method of squaring and sawing a board
- construct a chicken house for twenty-five hens
- demonstrate the proper procedure for gas welding two pieces of iron or steel together
- prepare a garden and properly plant seed
- match gear selection and engine's gear speed with load on a tractor

- manage the brood in an apiary
- prepare a broiler house for baby chicks
- keep feed records and records of expenses and receipts in a pig operation

It is important to note that each of these competencies is expressed in terms of performance-based objectives.

The use of performance-based objectives requires that objectives be defined in such a way that they state what is expected of the trainee for him to demonstrate that he has mastered the competency. It is not enough to say that the student should know about milking or understand how to graft a fruit tree. The objective must be stated in terms of an active verb.

The Oklahoma Department of Vocational and Technical Education at Stillwater lists performance terms and their synonyms, which are used to define performance-based objectives. These terms and their synonyms include such verbs as *name, identify, describe, order, distinguish, construct,* and *demonstrate.*

The process of teaching competencies involves more than stating the objectives in performance-based terms. Professors John R. Crunkilton and Alfred H. Krebs have developed a formula for selecting teaching techniques and instructional aids:[2]

lesson topic + behavioral objective + factors = teaching techniques + instructional aids

According to this formula, once the teacher has decided on the lesson topic based on the overall plan, he should clarify the course objectives and identify any specific factors, such as the abilities and interests of the group, the size of the class, and the teacher's own competencies, that would affect the teaching circumstances. Once the instructor has analyzed these elements, he is in a better position to determine the teaching techniques and instructional aids that he can use. Professor Crunkilton, who spent six months at the Farm School on a Fulbright grant, was able to inspire the teaching staff to review these techniques and aids in relation to the school's program. Unfortunately, recognizing the need is only the first step in a process that requires considerable time, effort, and financial support; of equal importance was his emphasis on teaching agriculture through problem solving.[3]

Professors Crunkilton and Krebs listed the following variety of teaching techniques in their book:

- panel discussions
- experiments
- programmed instruction
- team teaching
- micro teaching
- videotapes
- individualized instruction
- seminars
- debates
- questioning
- field trips
- brainstorming
- educational TV
- testing
- contests
- games
- discussions
- supervised study
- student reports
- role playing
- demonstrations
- resource persons

Many of these techniques have never been used by Farm School teachers, and a few are unknown in some Greek schools.

ANALYZE NEEDS

Instructors find preparing outlines for vocational agriculture courses time consuming, particularly as they must decide what skills are needed and which should be emphasized. For many years the Farm School used an apprenticeship approach in which the students spent half a day in the classroom and the other half in one of the practical departments, working on a project.

This approach for hands-on instruction made it difficult to keep a balance between production and education. In a school these two sectors do not mix easily. If students are required to do productive work after they have acquired a skill, they feel that they are being exploited. Once students have learned to hoe a field, they are not interested in additional hoeing to produce a crop for the school's profit. One way of overcoming this problem is to use a card listing the competencies required of a certain student. By this system the student and the school have a place to keep track of the skills a student has mastered.

Many years ago a vocational teacher analyzed the Farm School's skills training in a very effective way. He first spread a piece of wrapping paper about ten meters long around the room. On this he listed the various occupations in which graduates might be employed. He encouraged the staff to list the skills needed for each occupation and under each skill to list the activities required to develop it. Under each activity he enumerated projects that the students could work on to acquire the skills. For example, in carpentry he suggested a beehive, a backgammon set, a wooden suitcase, or a chicken feeder. He emphasized the importance of each student selecting his own project.

Students should be involved in selecting the skills that they acquire in any particular course. It is also important to determine whether or not the students are actually learning the basic skills listed. To assist in this process the school has adapted a series of skills charts prepared by a vocational school in England.

On the bulletin board of every laboratory is a chart, similar to the one illustrated, for each course. At the beginning of each term the teacher helps the students list the competencies that they should learn during the year. As students acquire a skill they put a star opposite their names indicating that they have successfully completed this part of the training. In the English system the number of checks under each skill indicates the level of proficiency; in the Farm School instructors use only one check to avoid conflict with the students. Although instructors in the Farm School have not universally accepted the skills charts, more teachers are using them each year. Skills charts have also been helpful in verifying which of the three hundred and seventy-five skills are being taught and in which laboratory.

One of the difficulties teachers encounter in vocational agricultural training programs when trying to organize the teaching of competencies is the dearth of teaching materials. To the outsider, selecting a number of competencies, listing them on the skills chart, and proceeding to teach them to students would seem like a relatively easy task. Unfortunately, it is not such a simple process. Professor Curtis R. Finch and Professor John R. Crunkilton, experts in curriculum development, listed eighteen stages involved in developing curriculum materials.[4] The Farm School has failed to prepare adequate materials primarily because neither the funds nor the staff needed to devote full time to the preparation process has been available. A useful tool for teaching competencies—the Learning Activity Package (LAP)—was introduced to the Farm School by Professor Crunkilton. Much of the material provided in the LAPs, which have been developed under the auspices of the Interstate Distributive Education Curriculum Consortium, can easily be translated and applied to teaching competencies in a developing society. As compared to the teaching packages described in Chapter 10, which require an enormous effort to prepare, a LAP is a brief, two- or three-page document that can be adapted to changing geographic or cultural situations.

EMPHASIZE TECHNICAL SKILLS

A shortcoming in agricultural training programs has been the emphasis on agriculture at the expense of related technical skills such as carpentry, masonry, machine shop work, and electrical repair. Charles House made a major contribution to the Farm School by insisting on technical

ACADEMIC DIVISION

DEPARTMENT: MECHANIZED AGRICULTURE
CLASS: First year Technical School
LESSON: Greenhouse Construction
HOURS/WEEK: THEORETICAL 2 PRACTICAL 2

STUDENTS' NAMES	SKILLS TO BE LEARNED												
	Construct frame of greenhouse	Cover greenhouse with plastic	Select material for frame and cover	Construct hotbed (manure heated)	Make soil mixtures (sandpit manure)	Sterilize Soil	Prepare soil for planting	Sow seeds in hotbed	Identify seeds of flowers and vegetables	Care for growing plants in hotbeds	Sterilize seed coat before planting	Plant flowers and green hou	
KIDSELAKIS, Stamatis	+	+	+	+	+	+	+	+	+	+	+	+	
KOMKIS, Pantelis	+	+	+		+	+	+	+	+	+	+	+	
KOPARANIDIS, Zisis	+	+	+		+	+	+	+	+	+	+	+	
KRIAPIS, Agorastos	+	+	+	+	+	+	+	+	+	+	+	+	
MANTZANAS, Dimitrios	+	+		+			+			+	+	+	
MAYROGEORGOS, Theod.	+	+	+	+			+						

training for the students. Such skills are invaluable to farmers, who cannot afford to hire a technician for every job that needs to be done. Many graduates have become sufficiently efficient as plumbers, electricians, or mechanics to supplement their farm income working in these trades.

When the students begin the course in farm construction at the school, they have a choice of building a small piggery, a chicken house, or a calf barn to use in their student projects the following year. Once they have chosen their building, the students are responsible for every aspect of construction. In designing the structure they become familiar with various types of farm buildings and learn the importance of preparing a bill of materials and a budget. They make concrete blocks, lay out the site, and learn to square a building. As the students progress in the construction, they seek the teacher's advice: a far more effective approach than the teacher giving a series of theoretical lectures on farm construction.

At the Indian Meridian School in Oklahoma the students actually built a new residence every year as the basis for their training course in construction. They designed the house and built it from the foundations to completion. At the end of the school year they sold it and used the money to invest in materials for another house and to buy equipment for the school.

To provide an opportunity for the students to practice their skills, the school requires them to undertake a home improvement project during vacations. They paint woodwork, build toilets, install water lines, or construct a cement path. This project demonstrates to their parents that the purpose of the Farm School education is to encourage students to implement the innovations that they have learned. For many years, at graduation the students were presented with a tool kit provided by the Cooperative for American Relief to Everywhere (CARE). The kits were intended to motivate the graduates to continue making changes from the time that they first returned home.

The School's faculty members receive no greater reward than that of visiting their former students and discovering the extent and variety of the impact they have had on their villages, particularly in applying their skills. One graduate, Christos Ouzounis, supplemented his farm income by doing plumbing work for others and eventually became a successful plumbing contractor in his town. Another, Garifalos Karay-iannis, built his own house. He proudly told a visitor that he had done all the iron work, the plumbing, the electrical installations, the painting, and the masonry. Michael Lialias, who was president of his village, built its first trench silo, modeled after one he had seen at the school. Other graduates have excelled as hog raisers, truck farmers, and dairy technicians. When asked what makes Farm School graduates different, outsiders have given two general responses: that they know how to work and that they have a "certain spirit" about them. In discussing a graduate, a Ministry of Agriculture inspector remarked: "If Greece had three such graduates in every prefecture it would be a very different country."

TEACH VILLAGE CRAFTS

In the postwar period after many villages had been destroyed, the Greek government organized itinerant craft schools. Skilled carpenters and masons from the villages were given crash courses in how to teach. They then spent three-month periods in central villages teaching their skills to young boys. In each course the class built a permanent structure for the village. When they completed the course students received a certificate as carpenters or masons. Many of the finest homes in rural Greece today were built by the graduates of these itinerant schools.

For many years the Farm School operated a craft center to teach weaving, dyeing, embroidery, knitting, macramé and other related skills, but it was difficult for the staff to decide on the aims of the center. Experience showed that teaching individual crafts as a vocation required major emphasis on dexterity and production. Products were not commercially viable unless the trainee learned to produce quantity as well as quality. On one hand, crafts produced for the market demanded long hours of drudgery with little reward. On the other hand, crafts produced for self-expression and recreation provided tremendous satisfaction but were seldom financially rewarding, except when the artist was particularly talented. The founders must clarify their goals before starting a craft center. Because the school failed to define its objectives, the craft center was never really a success.

After closing the center, the Farm School organized a new program to encourage self-expression for the students through crafts rather than to teach them how to produce commercially. The aims of this new program were defined by Elizabeth Holdeman, the instructor:

- To elicit and develop the innate creative potential of the student.
- To sharpen the student's visual sense and educate the aesthetic taste.
- To teach the students to express these faculties through such crafts as weaving, embroidery, macramé, drawing, and collage.
- To give the student the satisfaction of mastering a technique and the joy of creating through the coordination of mind, heart, and hands.

The National Foundation and the Ministry of Agriculture have been far more successful than the Farm School in organizing craft production. They have set up short courses and in-service training programs for older women who are already producing handicrafts. These women have already demonstrated their willingness to meet the rigorous demands of turning out crafts for the home market. By learning to adapt their products to Western tastes, village women can make crafts that rely on traditional designs, for which there is a growing market.

Aristotle used a sentence that reinforces the importance of competency training: "The things which we are to do when we have learnt them, we learn by doing them."[5] Lee Myer, a member of the school's staff, taught the students that any practical task performed by one person can be learned by another once the latter has developed the necessary skills. This philosophy is extremely important for the peasants, who lack confidence in their own ability to learn technical skills. Another staff member, Demetri Hadjis, used to ask, "What good is it to have practical knowledge in your head if it can't find visible expression through your fingers?"

When the school was looking for someone to develop teaching packages, it consulted Dr. Francis C. Byrnes of the International Agricultural Development Service (IADS). IADS maintains an extensive file of people interested in working overseas. Dr. Byrnes has divided these into two categories—those with expertise and those who combine expertise with experience. He distinguishes between the worker who acquired skills through the experiences in his personal background and the worker who may be a capable teacher or development worker but who lacks hands-on experience. Both are valuable, but it is important to select the person most suitable for any particular job. The most successful development worker is the rare one who is able to combine the teaching experience and the practical background.

Several teachers have sought assistance from the Farm School because they felt that they lacked some sort of practical experience. On one occasion a vocational agriculture teacher from a Greek public high school was anxious to work in the practical departments of the school to acquire skills that he needed for his farm shop courses. On another occasion a university graduate who asked to spend ten days working in the field crops department of the Farm School said he had been appointed to teach farm machinery at a junior college but did not know how to drive a tractor.

One of the most unusual courses run at the Farm School was that for Greek village priests in the late 1950s. In Greece at that time the village priest was expected to supplement his salary by agricultural work. A group of priests in long, black robes and stove-pipe hats came to study beekeeping because they regarded this as an appropriate occupation for a priest. The staff took great pleasure in watching the bearded priests peering over the hives as they pulled out frames to find the queen bee. When they arrived they were interested in learning the theory of beekeeping; by the time they left they felt sufficiently confident about handling bees to want to buy a few hives and set themselves up in business.

Teaching both agricultural and technical competencies is a fundamental aspect of training farmers for development. The student contributes to the learning process by following the principle, "I do, I understand." The instructors must first define the competencies required of the student, express them in terms of performance-based objectives, and then determine the experience through which the student will acquire the skills. An effective competency training program requires clearly defined objectives, an instructor with practical experience, and an appropriate series of project activities to establish the trainee's confidence to deal with the practical problems facing peasant families in development.

One day a big argument split the village into two groups. They called Hodja to resolve the dispute, but his wife warned him that they might turn on him. As Hodja felt a responsibility he could not shirk, he put on his robes of office and set off with his wife to the square, where the villagers had gathered on opposite sides. The leader and a chorus of voices from the first group shouted to him to make sure that he understood their point of view. After listening awhile he stopped them and said, "Hey, you are right," and set off across the square with his wife trailing behind him. The second group shook their fists to convince him of the validity of their point of view. He listened and finally replied, "Hey, you are right, too." His wife pulled on his robes from behind and whispered that they could not both be right. He turned to her and said, "Hey, wife, you are right, too."

12

Changing Attitudes

- Is it possible to organize short courses that distinguish between the technical, financial, and social concerns of the trainees?
- What should be the major considerations in choosing course content?
- Is it advisable to integrate men and women in training programs?

Greek peasants have always felt inferior in comparison to farmers in Western Europe. They admire the organization and the meticulous attention to detail of their counterparts in the West. A number of Farm School staff members and officials from the Ministry of Agriculture have visited Holland on several occasions to study the holistic approach used by agricultural training centers there. They always return convinced that rural development requires much more than teaching competencies and management. Discussions at centers like Wageningen, the Volk High School at Bergen, and the Agricultural High School at Emmeloord have led them to understand the vital role of the social subsystem in agricultural training and development.

Dutch approaches to agricultural education have changed over the past hundred years.[1] Educators used to emphasize farm visits, but these have been replaced by short courses. Every farmer has a car, and few live more than two and a half hours from Utrecht, a centrally located city. Regional courses are held in schools, hotels, and agricultural schools, which can be reached by bicycle, and classes are usually spread over successive weeks. There has also been a movement from training at the level of secondary vocational high school to adult education.

Emmeloord is a new town built on a recently drained polder—land that was under water forty years ago. Professor Johannes Dopmeyer, the director of the agricultural department in the high school, and his staff are true philosophers of human development. He speaks of the challenge put to them by the local farmers in the early years of adult short courses. "You have turned us into Tarzans," they said. "You taught us to be superb technicians and first-rate managers, but you have not taught us how to deal with people." When he asked them what they meant, they replied, "We can repair our machines, rebuild our houses, produce the best milk in the world and maximize our profits, but we know nothing about how to relate to our wives, our children, our employees, our neighbors, our fellow cooperative members, or even ourselves. We are like Tarzans in a new jungle of over-civilization." These comments led these educators to establish a whole new philosophy of training for their short-course center.

THREE ASPECTS OF FARMING LIFE

According to Professor Dopmeyer farming in a family enterprise involves three aspects:

1. The technical subsystem—working with crops, livestock, machinery, physical materials.

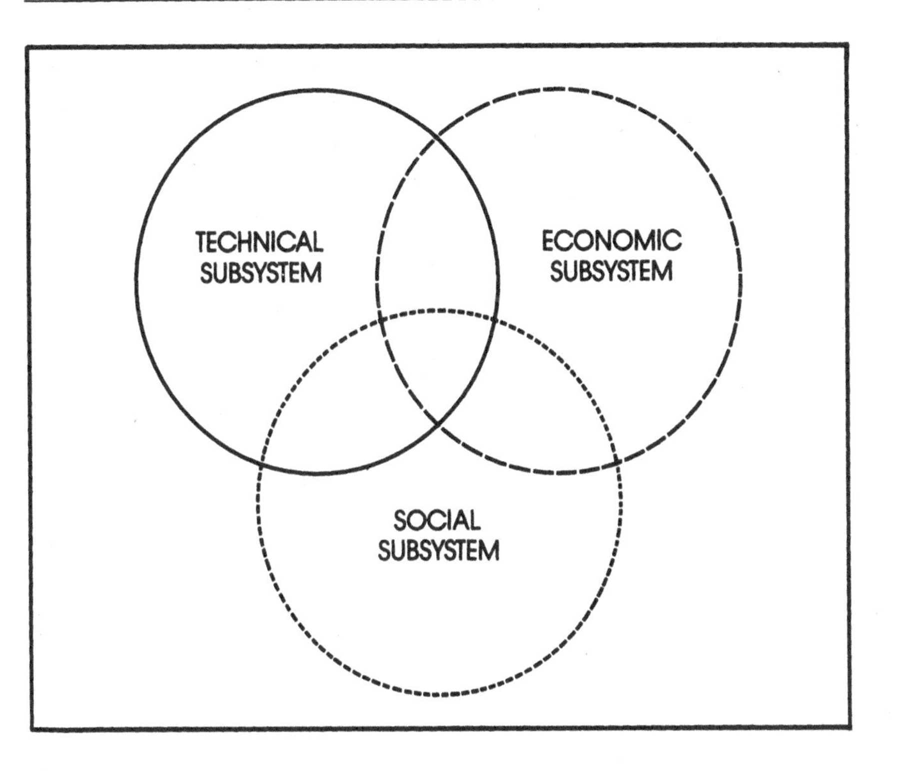

2. The economic subsystem—budgeting, financial planning, and record-keeping.
3. The social subsystem—the family's relationship with each other and with neighbors, partners, laborers, and cooperative members.

These three subsystems overlap and are interdependent, as shown in the diagram.

The educational system developed at Emmeloord is based on studies of farm families. In short courses (limited to twenty trainees) adults learn as much from each other as from the instructor. Instructors are the facilitators of discussions rather than lecturers. It has been found that adults learn better through practical experience than by listening to long lectures. The courses grow out of their own practical problems, so that the trainees can actually apply what they learn while the course is in progress. The concept of teachable moments is essential to effective teaching. Instructors point out that you do not teach ice skating in the middle of summer. Although Professor Dopmeyer has been designing these courses for thirty-five years, he feels that the courses are still in the process of development. Some trainees who were in their twenties

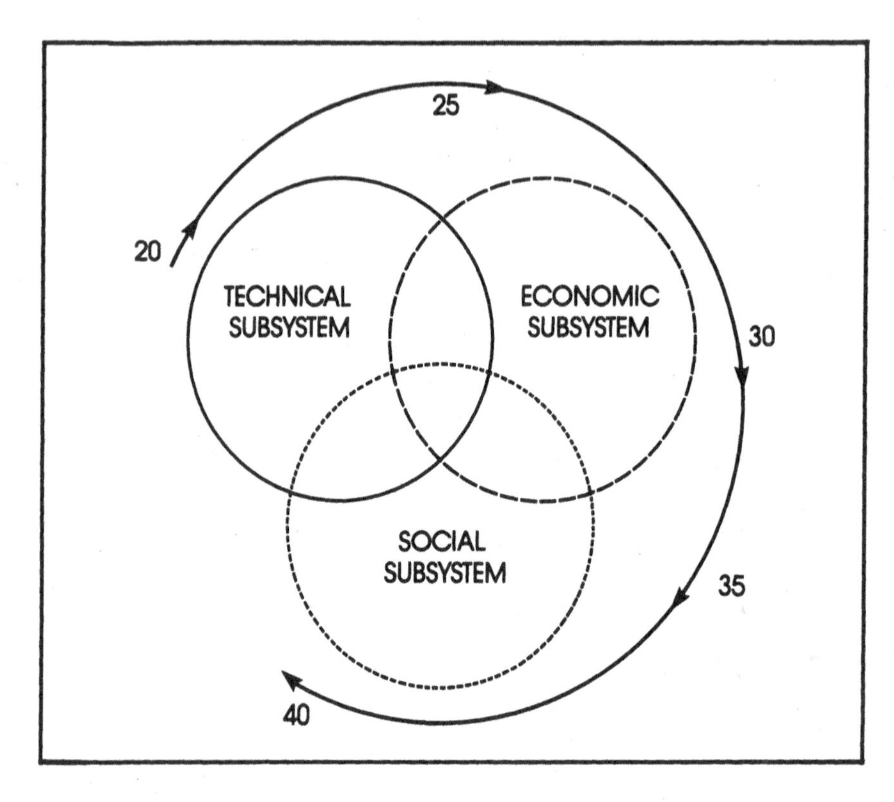

when he began are now over fifty—emphasizing the need for continuity in adult education from year to year.

As is shown in the diagram, the concerns of farmers tend to change as they grow older. In the early years (twenty to twenty-five years of age) the farmer's predominant interest is the technical subsystem. Young people do not pay much attention to either management or the social aspects of farm life. They want to know how to take tractors apart and put them together, how to cultivate fields, how to increase a cow's production, how to cook better meals, how to care for chickens—anything relating to farm production. As they develop a greater interest in management, they put more emphasis on the interrelationship of the technical aspects of agriculture. Trainees also become more concerned with the economic aspects of farming. At Emmeloord they participate in a one-hundred-and-twenty-hour course in management principles and practices. This course meets weekly and includes a maximum of sixteen trainees.

For farmers over thirty years, the school at Emmeloord offers a thirteen-day course for sixteen participants in social competencies for farming. This course was developed after the local farmers complained that they were being trained to be Tarzans. It is based on a discussion of the trainees' primary social needs, which include

1. How to discuss.
2. How to lead an organization, an enterprise, and an informal group.
3. How to cooperate with colleagues, with laborers, and with family (especially father-son relationships).
4. How to obtain information through observation, through reading, through asking, and through listening.
5. How to make decisions.
6. How to think.
7. How to negotiate.
8. How to handle conflict situations.

INTEGRATION OF TECHNICAL, ECONOMIC, AND SOCIAL ELEMENTS

There is an almost universal reaction in developing countries to the concept of using integrated technical, economic, and social subsystems for different age groups. Development workers feel that such an approach might be suitable for educated people but would be inappropriate for uneducated peasants. They also point out that social training courses will not work in developing countries because the people do not know their needs. Nothing could be farther from the truth.

Although educators may have to work more slowly to develop understanding and particularly to cultivate a close working relationship and confidence between themselves and the peasant, the Dutch approach used at Emmeloord and at the Volk High School Training Center for young farmers in Bergen is ideally suited to any society. Some might question the cost of such courses. Can a country like Tanzania afford courses in which only twelve people participate? Yet courses in social competencies for farm families are even more important in a developing country than they are in Holland. The level of training in these courses should be related to the wisdom of the trainees and not to their level of education.

The Modern Agricultural Enterprise Course at Emmeloord for farmers over thirty-five attempts to integrate the technical, economic, and social elements of the three subsystems. The thirteen-day course, which is divided into four periods, is spread over approximately two months. Period A deals with the various aspects of learning and decisionmaking. Period B covers basic social skills—listening, cooperating, communicating, and relating to others on the farm and in rural areas.

In Period C, the longest section of the course, the experience of the first five days is related to the trainees' home situations. The farm of one group member is used as a case study in a management game. The group, working in two teams, visits the farm, and each team prepares a five-year plan for its technical, economic, and social development. In Period D subjects for discussion tabled during the previous days are reviewed. At this point particular emphasis is placed on conflict resolution—recognizing, building, handling, and resolving conflict—by inducing conflict situations within the group and assisting participants in resolving the differences. This aspect of the training is somewhat similar to encounter training programs in the United States.

Professor Dopmeyer and his staff suggested some guidelines for the Modern Agricultural Enterprise Course. Two teachers are required—one to lead, the other to support and advise the leader. The course cannot be organized without the participation of the trainees themselves because the farmers' own experiences provide the basis for the course. Any group of farmers always knows more than one teacher: The teacher's function is to act as a catalyst to encourage the farmers to teach one another.

Franz Vader is the director of the Bergen courses offered at a special Volk High School Training Center for young Dutch farmers. These courses are prepared by the leader in conjunction with the participants. Each group of twenty to twenty-five trainees has an opportunity to participate in the development of the subject matter of their course. Primary emphasis is on the trainees' social problems rather than their technical or economic concerns.

The course goals are clarified in advance within the general framework of the centers' policy so that they are clear to the trainees on arrival. The courses are not specifically in agriculture but are intended for people involved in such areas of agriculture as managing farmers' cooperatives; rural development; agricultural policy; industrialization in agriculture; social aspects of farming, such as farm inheritance; "What goes on around us?"; and the problems of those leaving agriculture. The underlying principle of the courses is to start from where the students are and to build out of their own needs so that the participants do not feel that these are just general courses run for them by the center.

TECHNIQUES FOR SETTING UP EFFECTIVE PROGRAMS

The staffs at both Emmeloord and Bergen have defined some basic concepts for operating training programs that relate to changing the trainees' attitudes. The higher the educational level of the trainees the more quickly they seem to attain the training objectives. Although this is not always the case, those working in a developing society should recognize that the instructor must move more slowly, work at different levels, and develop very different goals from those established at Emmeloord.

Integration of the Group

At these schools allowances are made for the individual disparities in background and intelligence when dividing the students into small groups. Everyone has different talents to contribute—some technical, some economic, and some social. The instructor must be careful not to split the group in half in such a way that the group of students with lesser capabilities is in one section and that with greater capabilities is in the other. Instead the students should be separated into groups with similar numbers of good, average, and poor students. It is equally important that the distance between the most and the least capable student is not too great. Those at the top of each group help to instruct the ones at the bottom. The effective group leader encourages all of them to contribute in their own ways.

Professor Dopmeyer indicated that the most effective approach is to form new groups each year; otherwise they tend to become ingrown and try to exclude new participants. It is reported that in a family course established by the Kellogg Foundation at Michigan State University, the families continued to maintain a close personal relationship long after the course ended.[2]

The Emmeloord school shares the universal problem of how to successfully educate both men and women in management courses. After taking experimental courses in which men and women participated together, the women complained that all the attention went to the men. The course leaders appeared to be oriented toward teaching the men and not entirely confident in communicating with women. Very often instructors have little experience in mixed groups, making it difficult for them to function effectively in such situations. Professor Dopmeyer recommended starting separate courses of men and women and then integrating them.

C. Presvelou, professor of home management at Wageningen, feels that it should be possible to find articulate wives who could participate in integrated courses with their husbands. She feels that emphasis should be placed on preparing the women in advance to participate in the discussions. She also suggests the possibility of operating courses for husbands and wives together but providing opportunities for them to meet separately, depending on the subject matter. She believes firmly in developing courses in long-term farm planning by family groups, in which five or six couples might participate.[3]

Pointers for an Effective Program

Franz Vader summarized his observations with a number of suggestions that should be considered by those developing a short-course training program.

1. Help the trainees gain confidence in the center, based on their personal experience.
2. Don't let the center oversell itself; let those who have attended courses speak for it.
3. Encourage those who attend the courses to influence the policy on type of courses, working methods, and subject matter.
4. Recognize that this kind of training program is completely different from normal educational programs in that discussions start from where the trainees are.
5. Help the trainees to feel that it is their course, as opposed to being told by a teacher, "This is what you should learn."
6. Emphasize that farming as a way of life, not just as a means of increasing productivity using the fewest people at the lowest possible cost.
7. Establish an advisory committee of staff, farmers, and others related to agriculture.
8. Make it clear that the center is not an organ of any power group but exists to meet the trainees' needs.
9. Ask the participants to evaluate each training program as a guide to future courses.
10. Be clear about the purpose and goals of each course so that the trainees know where they are going.

The Family in Home Economics

Ita Hartnett, the supervisor of farm and home management for County Cork, Ireland, has evolved an impressive philosophy of development.[4] Her ideas about effective rural economics and management programs deserve careful consideration. She believes that development workers should think of home economics as having an agricultural base rather than agriculture as having a home economics base. In this way the farm family integrates the economics of the home and farm based on the agricultural potential. Planning and management should interpenetrate every aspect of farming and family life rather than being thought of as separate functions.

The motivation and incentive for development come from the whole family, not from the man alone. The tendency has been to train the man, hoping that some of what he learns will be absorbed by his wife at home. In fact, the whole family benefits from training, and only through the whole family can it be adequately disseminated. Women as well as men are eager for training in short courses, if these are useful to them. Too much emphasis has been placed on stitching and stewing, rather than on the development of the total human being and the integration of women into management.

Ita Hartnett emphasizes her definition of management as doing "what you want with what you've got." Starting with this definition, management training requires instruction on both sides of the equation:

what you want = what you've got

The development worker must help the trainees decide what they want according to what they have or choose what more they want if they can find a way to increase what they have. On the other side of the equation, they should understand what they have in terms of capital and income and how they can increase these. This approach ties in closely with the teaching package, "Relationship of Family Goals to Changes in Farming Methods," developed at the Farm School by Dr. Harry Peirce (see chapter 5).

Ita Hartnett described the need for a team effort in preparing the course syllabus. She went on to say that it requires advanced orchestration and should involve those who are intimately acquainted with the problems and needs of the individual farmers, the farms, and the farm families. The advisors should be the "organizers of the learning experience," using other agencies and integrating their efforts into a total teaching package. They should think of themselves as "promoters of the development of people"—an approach that relates to the social subsystem described by Professor Dopmeyer in Holland. For this reason successful management courses require joint participation of husbands and wives, working either together or independently.

According to Ita Hartnett the danger of formal education in farming and agribusiness is that it places the total emphasis on the business aspect of agriculture and shows minimal concern for the land or the animals. Actually development involves two objectives that are equally vital: (1) raising the standard of living, which may be defined as the quantifiable changes such as refrigerators, washing machines, automobiles, dishwashers, tractors, and other consumer goods; and (2) improving the values of living, the qualitative aspects reflecting human relationships and personal aspirations. It is particularly important to work toward the latter in cultivating new attitudes and changing the feelings that farm families have about their way of life.

CHANGING STUDENT ATTITUDES

The early courses at the Farm School from 1946 to 1965 emphasized production in areas such as processing food, making clothing, operating a dairy, and farming crops or livestock. Since 1965 course content has increasingly emphasized management, particularly record keeping and accounting. As yet there has been little concern in these courses for the social aspects of rural life.

Helping peasants change their attitudes has proved to be one of the most difficult aspects of the development process. Considerable controversy has developed about how much of this change is the product of social evolution and what can be attributed to the conscious effort of development programs. A valid question arises as to whether development workers have the right to try to change another's values.

Such workers do have the responsibility to help the peasant learn to adapt to changing cultural patterns, accept suggestions from outsiders, and develop the flexibility needed in a dynamic society. They must cultivate the peasants' interest in learning, opening their hearts to others, and setting new qualitative standards for themselves, their families, and their community.

The instructors' basic task is to help the peasants who think of themselves as kakomiris, the ill-fated ones, to acquire the values and standards of the nikokiris, the master farmers. They accomplish this task not as teachers but, in the words of Dr. Dopmeyer, as catalysts, making it possible for the nikokiris to share their wisdom and experience with the kakomiris. The sooner development workers understand that this is their most vital role the more successful they will be.

It would have been inconceivable to try to imitate the Dutch model in Greece thirty-five years ago, just as it would be inappropriate to transfer the Greek approach in its entirety to less developed countries today. But with cultural adaptation, many of the principles related to changing attitudes in Holland can be applied to Greece. Similarly, aspects of the Greek experience can be equally useful to developing countries. In either case, the solution lies with helping the peasant and his family adjust to the inescapable dynamics of change in the world around them.

13

Problem Solving

- How does problem solving relate to development?
- Why is this skill important for the development worker and for the peasant?
- Can problem solving be taught in training programs?

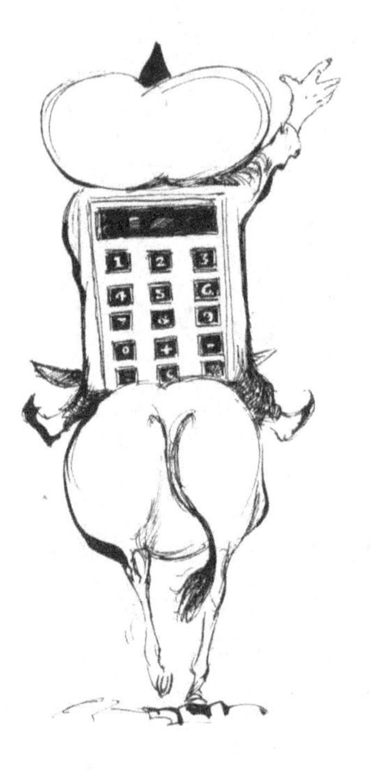

One day a man in Hodja's village died, leaving seventeen donkeys for his three sons. According to his will the oldest son would receive one-half of his donkeys, the second one-third, and the youngest one-ninth. When the sons were unable to divide the donkeys according to their father's wishes, they came to Hodja to resolve their differences. "You are fighting over nothing," said Hodja. "I will lend you my donkey and everything will be in order." Adding his donkey made the total eighteen, so that he gave one-half, or nine donkeys, to the eldest son; one-third, or six, to the second, and one-ninth, or two, to the youngest, making a total of seventeen. He bowed to the three young men, climbed onto his own donkey and headed for home.

Teaching problem solving is a process of expanding people's minds so that they will seek new solutions to old problems, try old solutions to new problems, and develop dynamic approaches to changing problems. In seeking to distribute the seventeen donkeys equitably among the three sons, Hodja first defined the problem: It was not how to divide them in accordance with the father's will but in a way that would make all the heirs happy. His next step was to collect the relevant facts: number of donkeys, number of sons, intentions of the father, and other such information. He then sought various ways of allocating the donkeys. Once he had listed all the possibilities in his mind, he evaluated each one and its implications, adding an extreme solution on the basis of a hunch. He selected the best solution, followed up on it, and was delighted that everyone was happy with the result.

In solving his problem Hodja followed the six basic steps that need to be taught in peasant societies:

1. Define the problem.
2. Collect information.
3. Seek alternatives.
4. Evaluate the implications of each.
5. Select the best alternative.
6. Follow up.

Many Western farmers might argue that these steps are so logical and simple to follow that they do not require any instruction. However, they are often remote from the pattern of thought of most peasants. Only after a thorough understanding of the way the peasants think is it possible to develop a technique to teach them problem solving.

BASIC ASSUMPTIONS

Before anyone starts to teach problem solving in a peasant society, he must make certain basic assumptions, which most Westerners take for granted. Edward de Bono, who is respected for his concept of lateral thinking, described a very successful course in thinking taught in a group of elementary schools in South America.[1] In Greece thirty years ago most university graduates were convinced that the peasant could not be taught how to solve his problems and that outsiders had to make decisions for him. "They can't read and write," and "They can't count to one hundred" were typical comments. Unfortunately, this feeling was shared by the peasants themselves.

Because peasants are inclined to use traditional patterns of thought, outsiders assume that they are poor problem solvers. City dwellers, especially in Third World countries, and very often the peasants themselves are convinced that peasants are less intelligent than their brothers in the city or their cousins in industrialized countries. As the Nobel Prize winning economist, Theodore Schultz, has said, "Farmers the world over, in dealing with costs, returns and risks, are calculating economic agents. Within their small, individual, allocative domain they are fine-tuning entrepreneurs, tuning so subtly that many experts fail to recognize how efficient they are."[2] With the proper motivation and training, peasants are not only interested in solving problems but quite capable of doing so.

Development workers assume that the peasant shares their assessment of the importance of particular problems. For instance, a planned parenthood organization was formed in Greece at the time family planning was considered the best solution to world hunger. But the peasants did not consider family planning a problem because they had long before discovered the use of condoms which were on sale at every kiosk or corner cigarette store. Furthermore, many more obstetricians were practicing in Greece than could be justified by the number of live births. Investigation proved that the doctors were making their money by performing abortions (especially for peasant women) rather than by delivering babies. The peasant had identified the problem and solved it long before the city experts arrived.

People eager to help peasants solve their problems in a hurry are apt to recommend solutions even before the problems arise. There is a fine line between leading and being too advanced. In 1937 the Farm School imported the first harvester-thresher to Northern Greece. Criticism of the machine was universal. How could it operate on small fields? It would never work in Greece where there was a surplus of labor. Nobody would be able to afford such expensive equipment. Yet immediately after World War II the Ministry of Agriculture organized short courses at the school to train technicians in the operation and maintenance of combines. In contrast to the success of this project, the first imported cotton-picker was exhibited by the school at the Thessaloniki Trade Fair in 1958, and farmers expressed only casual interest in the machine. However, six years later when there was a genuine labor shortage, cotton-pickers were imported on a large scale.

Using terminology foreign to the peasant intimidates him. Such terms as *systems analysis* and *computer technology* impress but frighten the uneducated villager and make him less receptive to innovative solutions. A World Health Organization group in Africa is reported to have explained to a group of villagers that inside an unwashed glass are multitudes of invisible germs that cause diseases. Much to the surprise of the experts, the villagers said they understood. "The missionaries have been teaching us for a long time about the invisible Holy Spirit of the Baby Jesus," they replied. If the Holy Spirit was present but invisible, why not the germs inside the glass?

Some rural problems can be solved intuitively; others require mathematical formulas or many hours of written work, both of which are methods foreign to peasants. Years ago a common practice in the Greek villages was to feed straw to dairy animals without any supplements, which reduced productivity. Experts published complex feed formulas that were seldom used because the peasants did not understand them. One agriculturist persuaded a few leading dairy farmers to try alfalfa hay and an occasional handful of grain. The increased milk production and profits soon convinced the farmers that his suggestion was worth following.

For many years Greek peasants felt that the advice they were receiving from agriculturists was theoretical, impractical, or ineffective. "The test of the benches of Krimni" is an expression used in a Macedonian village that has stone benches at the entrance. For years a cripple, a deaf mute, and two shrewd old men passed the time of day greeting new arrivals. After the four had sized up the visitors, the whole village soon knew whether the strangers were to be trusted. In another part of Greece Barba Pano enjoyed testing young agriculturists in his barley and wheat fields in early spring. While they were in the barley field, he would discuss wheat with them, and he would talk about barley in the wheat fields, to see if they knew the difference.

Young agriculturists and development experts new to the villages are often full of ideas before they know what supplies and equipment are available. The Farm School made this mistake thirty years ago when it first introduced a chisel plow, manure spreaders, and a forage harvester. Farmers asked where they could find this equipment but were told that it was not available in Greece. In contrast to this, when the school started a group of graduates in the broiler business with a grant from the Rockefeller Foundation, the staff imported parent stock to produce the chicks, prepared feed in the school mill, and built a model broiler house from local materials with student labor. In each case the chicks and the feed were sold to the graduates with credit from the Agricultural Bank. By having the chicks and the feed available the graduates were assured of success.

DEFINING THE PROBLEM

Probably the greatest difficulty in teaching people to define a problem is that they become so preoccupied with the symptoms that they lose sight of the problem itself. For years Greek agriculturists concentrated on increasing the production of peaches. The government had to bury several thousand tons of surplus peaches one year before it was realized that the problem was not quantity but quality control and the right varieties. The market needed peaches that ripened at different times and could be produced for canning and export. These requirements also apply to other crops like grapes, tomatoes, and apples.

The best way to help peasants and development workers define their problems is by encouraging them to think in terms of the journalists' five basic questions: who, when, where, how, and why? The reason for this approach was well illustrated by a foreign agricultural attaché who visited a remote dairy village and was shocked to discover that the milk delivered to the cooperatives was dirty and watered down. "They have to improve the quality of their milk" was all he said, instead of helping them understand their problem. He failed to state who should improve the milk, to establish some type of deadline, and to clarify where the problem existed, how it would be resolved, and why it was important. If they had had a precise definition of their problem, the dairy managers would have been able to start working on a solution.

In Chapter 2 the factors that hindered development in rural Greece were briefly reviewed. These factors are indicative of the peasants' difficulties in attempting to define a problem. For instance, how can they modify their deep-seated belief in the evil eye as a source of many difficulties? How can they be persuaded to accept responsibility rather than to shift the blame from themselves to anyone else they can find? Like most people, the peasant is sometimes his own worst enemy, and it is hard for him to realize that his own attitudes cause the problem. He finds it difficult to be objective when he is so personally involved.

Limited literacy is a handicap for many peasants because it leads development workers to assume that they are incapable of defining their problems. However, the school's experience with two particular peasants disproves this theory. The first was Katina, who though illiterate was without doubt one of the finest cooks the school has ever had. She developed her own system for keeping recipes by drawing lengths of lines in different places on a piece of paper to indicate quantities and ingredients. The other was Barba Pano, who could not read or write his own name. When he took a visiting teacher to see how he sold his lambs and was asked how he knew which lamb belonged to which ewe, he replied, "Don't you know each of your students and which family they belong to?" If Katina and Barba Pano were able to define their problems and solve them, there is no reason why others cannot be taught to do the same.

The difficulty usually lies with instructors who lack training in problem solving rather than the illiteracy of the peasants. Agriculturists are often embarrassed that they lack training in even the simpler skills such as driving a tractor or mixing cement, much less problem solving. One effective method used at the Farm School to overcome such embarrassment has been to set up special courses limited to agriculturists, in which they are encouraged to put on overalls and work with their hands. They usually look forward to returning to the villages to demonstrate their newly acquired skills.

A similar approach is needed to help instructors learn to teach problem solving. Agriculturists must learn to solve problems in an environment in which they will not feel threatened if they make mistakes. However, it is difficult to convince the authorities in a central ministry that a course in problem solving is a justifiable and necessary use of government money and agriculturists' time.

Trainees learn how to define problems by being exposed to them regularly. In Chapter 3 the Farm School student projects are described as small agricultural enterprises managed by the students themselves. The students own the projects as they do in 4-H and FFA projects in the United States. They operate the enterprises and make the related management decisions. Whatever profit they earn goes into their own pockets. Instructors are encouraged to help the students identify problems and define them as soon as they appear.

In one project an enormous sow was raised by a group of students. They took her to the boar twice without success. When, after the third attempt, they saw no results they slaughtered her. Only then did they discover twelve piglets inside her. They had defined the problem incorrectly. The definition should have been, "How do we determine if the sow is pregnant?" not "How do we stop losing money on this sow?"

COLLECTING INFORMATION

Assembling the facts that relate to a particular problem appears to be a simple process; yet in a developing society it is complicated by ingrained atttudes. Peasants tend to take a superficial approach to dealing with their problems. The case of the students and their sow is one example: The sow had not produced piglets; therefore, she should be slaughtered. In another project the students' crop of wheat froze. Their immediate reaction was to plant a spring crop so that they would not lose money. When they planted watermelons in the spring, they were horrified to discover that all their plants withered. One factor that they had not considered was the residue of the herbicide left after the wheat was harvested, which limited the types of spring crops that could be planted.

In the same way that superficiality inhibits the collection of adequate information perceptual blocks often hinder the process. In his exciting book on problem solving, *Conceptual Blockbusting*, James L. Adams defined perceptual blocks as "obstacles that prevent the problem-solver from clearly perceiving either the problem itself or the information needed to solve the problem."[3] Some of the perceptual blocks that Professor Adams lists are also common in developing societies.

Stereotyping among peasants and development workers alike leads them to make decisions based on preconceptions. Farm School staff experienced such a block when they discussed the possibility of putting boys and girls in the same dormitory building. Most of them were convinced that the school would face such problems as promiscuity, criticism from the villagers, and a drop in enrollment. But most of the parents were accustomed to young people living in unsupervised proximity when they were students in lower high schools. The students themselves developed brother-sister relationships and were surprised to learn about the faculty's concern.

Failure to isolate the problem impedes the identification of imaginative solutions. The school was faced with a Greek law forbidding the use of basement areas for classrooms and laboratories, at a time when funds were lacking to build urgently needed replacements. Harry Theocharides, the school's resident engineer, suggested that the problem had been incorrectly defined and that a more imaginative alternative was available. Why not dig the earth away from the solid foundations and convert the basement into a ground floor? The building has become one of the most attractive on the campus. The faculty had defined the problem as "how can we raise money to build new classrooms and laboratories." Theocharides' definition was "how can we convert the existing basement to use it legally for classrooms."

Overdelimiting the problem restricts the variety of possible solutions. The challenge confronting Walter Packard, an irrigation engineer, when he came to Greece with the U.S. aid mission in 1948, was to drain the swamp lands. He extended the scope of the problem by considering how swamp lands could be converted into productive fields. He decided that rice was the answer, despite everyone's conviction that it could not be grown in these areas. He not only succeeded in growing rice, but because of his project Greece was converted from an importer of $14 million worth of rice to an exporter. He had been able to add the extra dimension to his solution because of his experience in other countries in which swamp land had been converted into productive fields. Grateful rice farmers in the village of Anthili placed a bust in Packard's memory in the village square.

Looking at problems from different points of view increases the range and variety of related information. Peasants tend to conclude that what has not been attempted before cannot be done. Evangelos Pantelakos, a Farm School graduate, owned a field near a river that was so sandy that nothing had ever been grown there. At the Farm School he learned that potatoes do well in sandy soil and decided to try to grow them. He dug a well, built a makeshift tractor out of his diesel pump, and forged cultivation tools in the local blacksmith's shop. All the villagers predicted that his project would fail, and even he had doubts about it. The first year he had a reasonable crop; the next year he had a bumper crop. Soon, all the other villagers who owned land by the river were growing potatoes, thanks to his willingness to approach the problem from a different point of view.

SEEKING ALTERNATIVES

The word *paradigm* comes from a Greek word *paradigma* (example). In *Powers of the Mind*, Adam Smith introduces the concept of the paradigm in problem solving.[4] Most individuals, and particularly peasants, are hesitant to break out of the area in which they have fenced themselves. In no area is this more obvious than in seeking alternative solutions to problems. The classic example of a limited paradigm is drawing four straight lines through nine dots (see p. 154) without lifting your pencil. The solution requires the solver to extend the lines beyond the paradigm of the imaginary square that the dots form. Once the lines are extended, the solution is easy.

A primary obstacle to development is the failure of academic education in developing countries to teach problem solving. Instead, it encourages conformity among the students. They must memorize the exact information presented by the teacher and regurgitate it in their examinations to obtain the highest grade. Teachers seldom encourage original thinking in their students or motivate them to innovate.

The Farm School organized a course for training center directors, which included a section on problem solving. They were presented with the nine-dot problem and one using six matches (p. 154), which they found even more complicated and frustrating. The directors were amused by the games but felt they were wasting their time; the only kind of education they had learned to respect was one that dealt with factual information.

EVALUATING THE IMPLICATIONS

The Farm School's experience has proved that a problem does not have a single, ideal solution. The objective is to seek the solution with the greatest possible benefit and the least negative impact. The more carefully the problem is defined the easier it is to evaluate the alternatives,

Draw four straight lines through the dots without lifting the pencil.

• • •

• • •

• • •

Take four whole and four half matches.

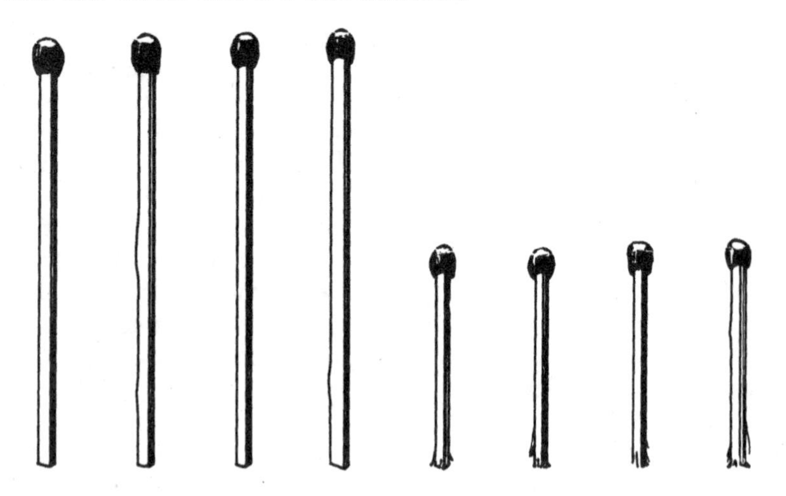

Arrange the matches so that they form three equal squares.

(The solutions to these problems are at the end of Chapter 13.)

though almost every alternative presents certain risks. These risks are what many peasants and the civil servants working in the villages try to avoid.

Based on his extensive observation of people from Western backgrounds, Dr. James L. Adams listed nine emotional blocks, which are remarkably similar to those of the Greek peasants thirty-five years ago. They include

1. Fear to make a mistake, to fail, to risk
2. Inability to tolerate ambiguity; overriding desires for security, order; "no appetite for chaos"
3. Preference for judging ideas rather than generating them
4. Inability to relax, incubate, and "sleep on it"
5. Lack of challenge; problem fails to engage interest
6. Excessive zeal; overmotivation to succeed quickly
7. Lack of access to areas of imagination
8. Lack of imaginative control
9. Inability to distinguish fact from fantasy[5]

Shame was and still is one of the most strongly entrenched emotional blocks in peasant societies. A common parental admonition in Greece is "Don't. It's shameful." Although this block is most easily observed among children, it is a dominant force in everyone's thinking. It relates to the fear of making a mistake because of the shame that failure might bring in the eyes of fellow villagers. A general comment made by extension agents about Farm School graduates is that they are eager to try new ideas. They seem to have overcome the apprehension associated with the shame of making wrong decisions.

Clearly peasants and urban dwellers share emotional blocks, but rural people are less sophisticated about hiding their weaknesses, which are therefore more apparent. These emotional blocks make it difficult for them to be objective in their evaluation of alternatives.

SELECTING THE BEST ALTERNATIVE

The fifth step in the problem-solving process is selecting the best alternative. Once a person has evaluated the implications, the selection of the best alternative should be simple. However, beyond the perceptual and emotional blocks that hinder the selection process a variety of cultural blocks stand in the way of the selection process in a peasant society.

Trusting a person who is not a relative or at least from the same village is as foreign to peasants as is cooperation. Time and again Farm

School students reiterate that it is a mistake to trust other people. A difficult problem in developing village economies grows out of this lack of trust: It has been one of the major roadblocks to organizing and managing effective cooperatives. The peasants all agree that cooperatives are the best solution to their marketing problems, but they seldom actually organize marketing cooperatives. The school has placed great emphasis on encouraging students to operate their projects cooperatively and to organize a cooperative store. But it takes time to break through the barrier and help them learn that they can solve their problems more effectively in an atmosphere of mutual trust.

Another significant block to selecting the best alternatives grows out of religious taboos, such as the church's initial attitude toward artificial insemination. When a station was established in Northern Greece in the postwar period, a young bishop was invited to give the customary blessing. His senior bishop, learning that he was planning to bless an artificial insemination center, was furious and forbade him to attend. Technical advances such as this are now generally accepted in Greece.

Cultural taboos also obstruct the problem-solving process. When the Farm School and the Ministry of Agriculture first organized a tractor-driving course for village women shortly after World War II, the immediate response from most of the peasants was that women could not learn to drive tractors, much less maintain them. Only the most progressive parents would allow their daughters to attend. The women were embarrassed when they were told to wear coveralls. Even the instructors had doubts about the program. The course had to be justified as a precaution in case of war, when the women would have to replace their husbands. Today tractor-driving courses for women are commonplace.

A major block to selecting the best alternative grows out of the innate desire of the villagers to please. They are eager to know which alternative is the most acceptable to the visiting official or agriculturist. They commonly ask questions like "What do you think?" "Which would you choose?" Development workers make a serious mistake when they answer such questions to prove their own problem-solving ability. They take a long time to realize that they are unwittingly depriving the peasants of the opportunity to learn to solve their own problems. Only extensive experience and practice will help the peasant overcome the ingrained attitudes that inhibit effective problem solving.

FOLLOW-UP

Following up on a choice is an extremely rare quality among most people. Once they make a decision, they tend to implement it without much consideration of the consequences. Careless behavior cost the

Farm School students their broiler stock. When a student responsible for raising broilers decided that the temperature was too cold for the baby chicks, he turned up the oil in the brooder and went to bed. Fortunately, the night guard discovered that the straw was on fire and extinguished the flames before the house burned down, but the three hundred baby chicks perished. Since then the students have learned how important it is to follow up their decisions.

Few people appraise the solution they select once it has been implemented. Ever since Greek farmers discovered forty years ago that fertilizers increased their crop production, they have applied them at high rates without checking the results. Since the fertilizers probably are not necessary in such large amounts, the peasants are wasting millions of drachmas annually. They feel that they found the solution and see no reason to question it.

The examples used in this chapter indicate how important problem solving is to development. Time and again Farm School staff have observed changes in the approaches of both students and graduates as they have become more proficient as problem solvers. Effectiveness in solving problems distingushes the private in agriculture from the sergeant, the kakomiris from the nikokiris. It is not uncommon to find peasants who have learned to solve problems from experience or by intuition. The ability is there; it remains for the development worker to help them learn how to use it.

Solutions to Problems on p. 154.

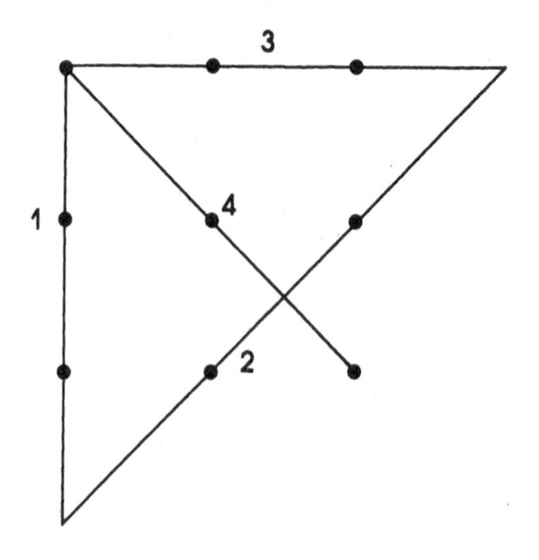

The essence of solving this problem is to break away from the paradigm and cross the matches.

1. Place the four whole matches in form of two Xs.

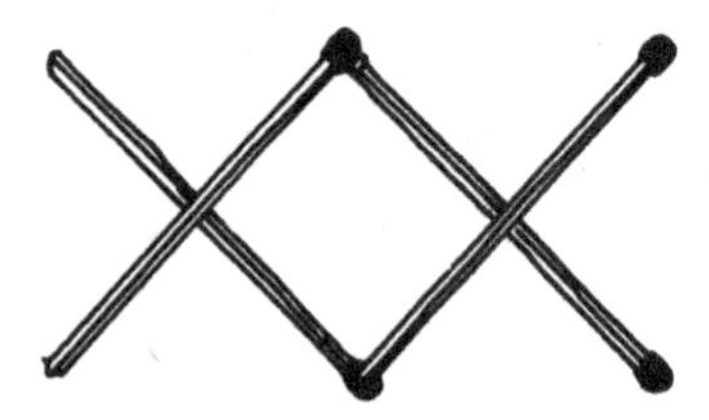

2. Place the four half matches at the ends.

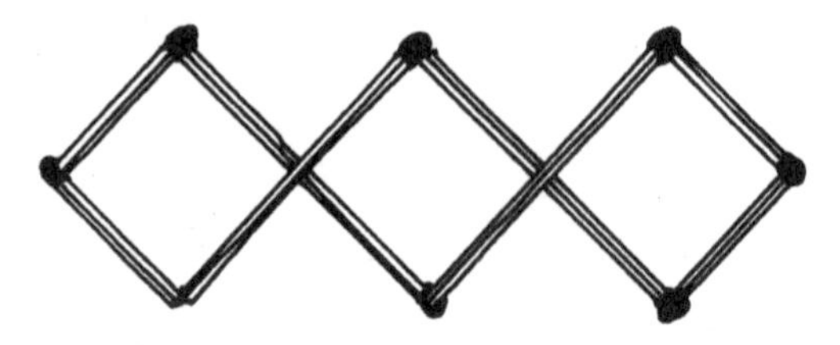

14

Building Self-esteem

- How can villagers be helped to understand themselves?
- Can development workers help them develop better self-images?
- Do training programs broaden their outlook?

One day Hodja's apprentice said, "Hodja, everyone says you're good. Does that mean that you really are good?" Hodja replied that this was not necessarily so. The boy then asked whether if everyone said Hodja was bad it would mean that he was bad, and again Hodja replied negatively. When the apprentice enquired how he could tell, Hodja told him that if the good people said he was good and the bad people that he was bad, then he was good. He paused for a moment, scratching his beard, and then continued, "But you know how hard it is to tell which are the good people and which are the bad."

Dr. John House founded the Farm School primarily to inspire self-esteem among village youth. He felt that the most important part of development was to help young people become aware of and accept themselves for what they were, by recognizing their shortcomings as well as their positive qualities. He was convinced that with guidance they could learn to overcome their negative feelings about themselves and reinforce the positive ones. In short, he felt that people could change. His goal was to encourage his students to understand their potential and identify themselves, in Hodja's phrase, with the "good" people. His goal was closely akin to the Socratic concept of "Know Thyself."

POSITIVE APPROACHES

To each new group of Farm School students the director describes the experience of a young sculptor who spent a summer at the Farm School working on a piece of marble from a nearby quarry. An old peasant woman watched with admiration as the visitor was completing her statue of the Madonna and Child. After studying it for a time the woman asked, "How did you know it was inside the marble?" The director finishes his story by saying, "One of your major efforts as students should be to better understand yourselves." The statue is a constant reminder that a primary goal of rural development is to help the peasant discover "what is inside."

The statue teaches another important lesson. As the sculptor was adding the final touches to her masterpiece, she accidentally broke off an arm, which was later replaced, leaving slight cracks. Students admiring the statue only notice the cracks when they are pointed out to them. The cracks are compared to real or imaginary flaws in their character, which they tend to magnify out of proportion. Trainees are encouraged to concentrate on the beauty of their personal "statue" rather than on occasional cracks barely visible to others. Emphasizing people's strengths rather than magnifying their weaknesses is as important as helping them discover the potential that lies within them.

Two hypothetical groups of peasants briefly described in Chapter 1 have opposite views on how they see themselves, their environment, and their fellow villagers. On one hand are the kakomiris, the ill-fated peasants who seem to dwell on their own weaknesses. Every project that they or their wives undertake in the home, on the farm, or in other activities seems to end in failure. The world views the members of this group as true losers. On the other hand are the nikokiris, the master farmers, who build on their strengths. They are the landlords in the sense that they are lords of their land, of their lives, and of their work. As a team they support each other and build confidence in themselves, in their enterprise, and in those around them. In the eyes of their neighbors they are winners.

In Chapter 1 the comparative list of the thought patterns and behavior of the kakomiris and the nikokiris illustrates the contrast between them. The examples are extreme, but they do provide a basis for understanding the outward manifestations of the way the members of each group see themselves and those around them. A major challenge to the development worker is to help unsuccessful peasants feel more positive about themselves.

Awareness of this disparity is important, because inspiring self-understanding and a positive feeling about one's self is an integral part of the process of building self-esteem. Development workers may talk of changing attitudes, teaching competencies, instilling knowledge, and organizing managerial skills, but the will to change rests with the people themselves. Development work begins with individuals and families and spreads to communities and regions. It encompasses the happiness or satisfaction of the whole human being, not merely his economic welfare. In a candlelight ceremony at the Farm School the students use a phrase attributed to Aristotle: "The one thing man seeks for in itself is happiness."[1] Economists are preoccupied with raising standards of living measured in terms of productivity, water supply, food, roads, and electrification. These may lead to material satisfaction, but without qualitative values of living reflecting human relationships and personal ambitions, development does not automatically bring happiness. It must be the outgrowth of individual aspirations.

The rural family is not only the beneficiary of development work but also a prime force in accelerating the process. In his Nobel Prize lecture, Theodore W. Schultz expressed the essential elements that should be a primer for anyone studying development:

> We have learned that agriculture in many low income countries has the potential economic capacity to produce enough food for the still growing population, and in so doing can improve significantly the income and welfare of poor people. The decisive factors of production in improving the welfare of poor people are not space, energy and cropland; the decisive factor is the improvement in population quality.
>
> . . . most observers overrate the economic importance of land and greatly underrate the importance of the quality of human agents.
>
> What many economists fail to understand is that poor people are no less concerned about improving their lot and that of their children than rich people are.

> A fundamental proposition documented by much recent research is that an integral part of the modernization of the economies of low income countries is *the decline in the economic importance of farmland and a rise in that of human capital—skills and knowledge.*[2] (Emphasis in original.)

Development workers must be convinced that people can and want to change their attitudes and behavior patterns. But emphasis must always be on what village families want and can do for themselves, rather than on what others can do for them. To help peasants make these changes those working with them must first understand how they feel about themselves and the world around them. At the heart of successful development is the recognition that people are not lamps to be filled but lights to be lit. The challenge of development is to discover the spark that ignites the flame, the spark that makes people want to find out "what is inside" and how they themselves can contribute to improving the quality of their lives.

TRANSACTIONAL ANALYSIS IN DEVELOPMENT

A tool that has been useful in helping Farm School students understand themselves, their families, and their communities is the theory of Transactional Analysis (T.A.), first developed by the psychiatrist, Eric Berne.[3] According to T.A. every person manifests three interrelated elements of his or her personality—the Parent, the Adult, and the Child—each of which has a definite behavior pattern. Dr. Berne's theory can help the development worker recognize that an individual's behavior can be directed by any of these aspects of the personality at a particular

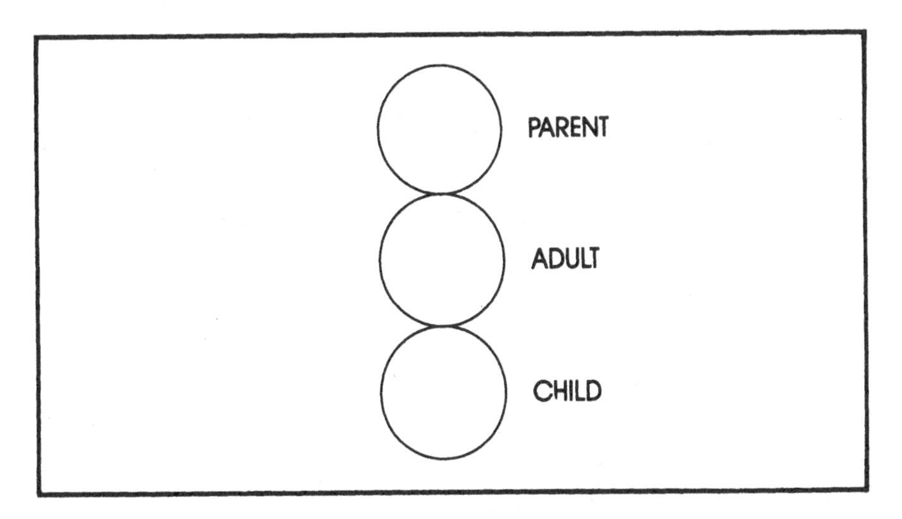

time. It is not possible in a brief chapter to present more than a very superficial summary of T.A. and its value as a tool in helping to interpret behavior patterns. Development workers should be encouraged to read such books as *Games People Play* by Berne, *Born to Win* by Muriel James and Dorothy Jongeward, and the bestseller, *I'm O.K., You're O.K.* by Thomas Harris.[4] Trainees from Third World countries studying rural development at Reading University in England are encouraged to take a course in transactional analysis.

Dr. Berne describes the Parent as that part of the personality that is learned by observing parents or other older people. The Adult attempts to understand facts, experiences, and relationships and to deal with them logically and unemotionally. The Child learns by reacting to its physical and social surroundings. According to Dr. Berne, a person's personality sectors evolve from experiences before the age of five. The Parent reacts automatically, using traditional responses. The Child tends to react emotionally, whereas the Adult tries to understand the circumstances and the interpersonal relationships involved.

All three personalities were manifested in a village woman when a development worker walked into her home moments after her child had broken a much-prized crystal candy bowl. The furious mother stopped screaming at the child when she saw the visitor. Her facial expression, body language, and whole behavior pattern changed instantaneously as she turned with a polite greeting. Three minutes later she was explaining how to make the liqueur she served. In the space of five minutes she had moved through three states of behavior: the emotional Child, the formal Parent, and the intelligent Adult.

Interpersonal Transactions

Identifying which personality aspect is dominant in another person can help in communication. If two people communicate from the same aspects of their personalities (Parent-Parent, Adult-Adult, Child-Child), the discourse will be parallel, or without conflict. Also, if the Parent in one talks to the Child in another, and the Child responds to the Parent, the conversation will be conflict-free. However, if the Parent of one person speaks to the Child of the other, who in turn responds from his Parent to the first person's Child, a crossed transaction takes place, and communication breaks down.

For example, when a development worker and a villager discuss a probem logically and unemotionally there is a parallel transaction—Adult to Adult. Likewise when the same two people condemn the national government or a neighboring country, they set up a parallel discourse between their Parents. Development workers often talk down from their Parent to the Child of the villager, who reacts from his Child back to the Parent of the development worker; this conversation is also parallel and conflict-free. But if the development worker addresses the Child of the villager from his own Parent in a critical manner and the villager becomes defensive, responding from his Parent to the development worker's Child, then there is a cross-transaction that breaks down communication.

Social Responses

Stroking (social intercourse) is a fundamental T.A. concept that is important for the development worker to understand. Just as "infants deprived of handling over a long period will tend to sink into an irreversible decline," so adults deprived of social recognition or contact become unhappy or deeply depressed. Dr. Berne regards a stroke as the fundamental unit of social action. More than anything people seek positive strokes: love, affection, compliments, encouragement, recognition, response. And if they cannot get them, they are willing to accept negative strokes as a substitute because they still represent a sort of social intercourse. The challenge for those involved in development is to establish the atmosphere that will yield the maximum number of positive responses.

The development worker must be aware of how people in a rural community spend their time. Through their interactions with other people, which may include rituals, pastimes, activities, games, or intimacy, they ensure that they will receive some type of response. Everyone's goal is to obtain as many positive responses as possible.

"Yassou" (hi), says the Greek peasant. "Ti kaneis" (how are you?) is the reply. By participating in such rituals, which involve agreed upon behavior, the peasant gives and receives socially acceptable responses. More than one development worker has failed in Greece because he did not understand or observe the importance of ritual in village society.

Greek villagers like to talk about crops, the weather, the bad merchants, lazy civil servants, the central government that neglects them, their livestock, the next village, and a neighboring country. A series of programmed exchanges between two people involving a single subject used to fill time is called a pastime. Gossip is one of the most common pastimes in the village.

Activity is a way of structuring time through work. It yields more responses because it engenders recognition from others. For men, plowing, planting, harvesting, tending the animals, or repairing the equipment are typical activities. Village women are admired for being busy, whether caring for their home, embroidering, knitting, or weaving. But increasingly they are seeking paid employment outside the home.

"Games" (subtle ways of getting attention) consume a large proportion of leisure time in peasant societies. Dr. Berne and his students have analyzed a variety of commonly played games used either to evoke responses or to reinforce already existing negative feelings. The titles for some of the games, "Kick me," "Poor me," "See what you made me do?" indicate the negative feelings they generate.[5] The attitudes of many urban inhabitants toward rural people intensify these feelings.

A typical game was witnessed by a development worker when a villager whom he had known for many years took him to dig up a discarded millstone on the land of a shepherd friend. Just as they had finished digging up the round stone the shepherd walked by with his flock. The intruders stopped all activity. There was a long period of silence as the shepherd and sheep walked past. From a fair distance away he turned and said, "You should have asked me, John. You should have asked me." When the shepherd was completely out of sight and hearing, John started cursing him in a violent manner, recalling all the favors he had done for him and assuring his foreign friend that he would find a way of getting even with him. The millstone was returned to its original place. According to Berne this game is called "Now I've got you, you SOB." The Greek peasant equivalent phrase is "Tha se kanoniso" (I'll fix you).

The most satisfying way of structuring time is through intimacy. According to Thomas Harris,

> A relationship of intimacy between two people may be thought of as existing independent of the first five ways of structuring: withdrawal, pastimes, activities, rituals and games. It is based on the acceptance by both people of the I'm OK—You're OK position. It rests, literally, in an accepting love where defensive time structuring is made unnecessary. Giving and sharing are spontaneous expressions of joy rather than responses to socially programmed rituals. Intimacy is a game-free relationship, since goals are not ulterior. Intimacy is made possible in a situation where the absence of fear makes possible the fullness of perception, where beauty can be seen apart from utility, where possessiveness is made unnecessary by the reality of possession.[6]

When this unity exists in a family it generates bonds of loyalty that outsiders find difficult to comprehend.

Fun is another satisfying social interaction.[7] Many villagers enjoy dancing, singing, and playing cards, soccer, and a variety of locally devised games using simple objects such as stones, sticks, and rope. *Kefi* is one Greek word for fun, but it has much broader implications. It has more spirit and implies a joy in living to the core of one's being.

On his way into a village a man stopped at a coffee shop for a rest. When he asked what kind of people lived in the village, the innkeeper enquired where he came from and what kind of people lived there. "I come from Kerasia up in the mountains. They are wonderful people up there," he replied, and was told that they were very good people in this village too. Six months later another traveler asked the same question and was given the same question as a reply. The traveler said that he was from Kerasia, but went on to say that the people in his village were rascals and would "take the gold out of your teeth." "Isn't that too bad," replied the innkeeper. "Those are just the kind of people you're going to find in this village."

One of the first tasks of the development worker should be to seek the equivalent term and its meaning in the language of the people with whom he works.

The way trainees spend their time in training centers, whether in rituals, pastimes, activities, games, or intimacy, determines the satisfaction they gain. A sense of intimacy among them inspires confidence and makes them eager to return for additional courses. Years of disappointment and poverty in peasant societies have made villagers hesitant to expose themselves to close relationships outside their family. They start protecting their children from intimacy at an early stage and prepare them for the harshest games in life. An important goal of development should be to promote gamefree relationships by generating genuine feelings of intimacy beyond the family.

When individuals do not obtain adequate responses from a group, they may withdraw emotionally or physically. Withdrawal is a rare way to spend time in the Greek village, where the word *privacy* does not exist. Widowed old women spend hours alone sitting on the front steps of their houses, apparently hoping for an occasional greeting from a passerby.

A life script is a plan for life or a projection about the future; it is usually instilled in a person by his parents and reinforced by the Parent in his personality. A negative script ("I can't succeed; I'm going nowhere; I'll always be a nobody") predominates among rural children, especially when they are not able to continue their education. People carry their script with them wherever they go, as illustrated by the story of the men from Kerasia. The Farm School staff is constantly seeking new approaches to help the students change their negative scripts to positive ones.

A trainee from abroad who spent a summer at the Farm School wrote, "I sensed such caring, such an admirable ability to appreciate in each of us what we had to offer, to bring out the best in everyone gathered there. It was not the Farm School's way to inspire change by rude awakenings. Instead the Farm School meant to nurture what was already good in a person, to bring about change by constant affirmation."

Westerners are inclined to project their own value systems on peasants in developing societies. How very wrong they can be! For example, a village woman in the Ogaden region in southern Ethiopia lived in a thatched beehive hut with her four children under the age of five and another on the way. Two calves, cooking utensils, and sleeping mats lay about on her mud floor, along with a few bits of clothing. Despite the poverty the mother was full of laughter and gaiety, and the mother and children shared a warm relationship. When a visitor asked his interpreter how they could be so happy when they were so poor he replied that they had no way of knowing that they were unhappy.

When trainees from Third World countries spend extended periods studying in industrial societies they are inclined to cultivate a new script, forgetting many of the frustrations and limitations of their own country, which causes intense cultural shock when they return. Happiness is indeed relative and seldom static. Commenting on this, Richard Shayo, the director of short-course training centers in Tanzania, said, "Unless you have seen a better way of living you always feel that your way is the best there is." He expressed the frustration of returning to the Third World after a time in the West when he described his own experience following his return to Tanzania after two years in the United States:

> I went to America. I learned to listen to beautiful music so I brought back a record player. But I stopped listening to the music. How can you listen to beautiful music when your world is full of problems? I bought a refrigerator to fill with meat and beer, but I couldn't afford the meat or the beer. I put in water to keep it cold, but when I got my electric bill I took the plug out—I realized these were dreams from America, dreams which I had to forget if I really wanted to serve my country.

In the United States Mr. Shayo had developed a new script for himself, but soon after his return the society in which he lived and the reality of conditions in his country seemed to reinforce the old script.

PROGRAMS THAT BUILD SELF-ESTEEM

Institutions and development programs are made up of programs directed toward all three sectors of a trainee's personality. The ultimate goal of those involved in managing training programs at an institution should be to balance these three aspects. The institution should give priority to the needs of the trainees, operate well-organized courses, keep firm but fair rules, and be tidy (thus appealing to the Parent). It should emphasize problem-solving approaches and have flexibility in its program (for the Adult), combined with periods for fun and leisure (for the Child). This kind of institution provides the maximum amount of social interaction for the trainees and the staff, and it encourages students to have good feelings about themselves and those who are part of the organization.[8]

The most important function of the development worker is to instill feelings of compassion and self-esteem in the people with whom he is working by exemplifying the former himself and inspiring the latter in others. At the Farm School the process of generating a sense of compassion is referred to as "reaching out," a concept described by Henri Nouwen.[9] He depicted three aspects of people's lives that bear on development programs: their feelings about themselves and about their fellow man and their spiritual values. In their relationship with themselves people must learn to reach out from a sense of loneliness to a feeling of solitude. Loneliness reflects a feeling of isolation, whether alone or in a group; solitude is an inner peace that many experience regardless of external circumstances. Training centers in which staff members appear indifferent or too busy to care generate a sense of loneliness, whereas others that nurture concern and warmth inculcate a satisfying feeling in solitude.

In referring to a person's reaction to others when they first meet, Nouwen spoke of the need to overcome possible hostility and consciously develop feelings of hospitality. On their first day at a training center most trainees tend to be apprehensive and slightly hostile, feeling threatened and wondering whether anyone is really concerned about them. Unfortunately, staff members often sense this animosity and react with antipathy. It is essential to cultivate a sense of hospitality, making everyone feel warmly welcomed.

In the final section of his book, Nouwen spoke of a person's preoccupation with material possessions at the expense of spiritual values. Training center administrators often attach more importance to buildings, furnishings, and teaching aids than they do to more intangible elements that contribute to an institution's general atmosphere. A comparative study of a few centers would quickly show how important it is for the staff members to reach out compassionately to the trainees to help them develop self-confidence and positive feelings.

A special privilege of a long, continuous association with peasant friends is the gift of their wisdom and their quiet self-confidence. In every village a few men and women manifest an inner peace—a solitude—as well as hospitality toward their fellow man. They are spoken of in their village as nikokiris, the winners who resolve problems dispassionately, who are often motivated by a strong cultural heritage, and who delight in the lighter moments of village life. The experienced development worker soon learns how much more these leaders have to teach than he or she can possibly share with them. The worker also sees the need to help the kakomiris learn to emulate the happier and more successful nikokiris by recognizing their own untapped potential.

15

Metamorphosis

- Is there a dimension to development work beyond management?
- What personal qualities in a development worker can inspire people?
- What stumbling blocks can a development worker anticipate?

One day the Hodja was resting at a crossroad on the edge of the village. A stranger stopped to ask for directions. When Hodja asked him which village he was heading for the stranger looked hesitant and said he was not really sure. "Then it doesn't really matter which road you take," said the Hodja, with the trace of a smile on his face.

The process of metamorphosis, defined as "change in appearance, character, circumstances,"[1] has been been a source of fascination down through the centuries. For most people it has a mystical as well as a physical aspect. Writers as diverse as Ovid[2] in the first century and Kafka[3] in the twentieth, as well as a variety of artists, have attempted to interpret the concept of transfiguration. Biologists think in terms of larvae spinning their cocoons and turning into butterflies.

Which of the many roads to development is more rewarding? Very few of those working in rural communities have really thought this question through. Agricultural experts speak in terms of per acre yields. Others press for industrialization. Population experts emphasize family planning. Economists stress scarce resources and factors of production. Ultimately the answer grows out of the Irish concept of helping peasants do what they want with what they've got. The quality of village life can best be improved by better management, but this is only a part of the total development process. Peasants must be helped in their search for new directions, while at the same time retaining pride in the traditions of their culture. Distinguishing between enduring values and those that can be discarded is one of the most challenging tasks of the development worker.

Most development workers have grown to realize that they do not have the right to impose their views. Their job is to encourage the villager to ask questions and seek answers that will help him decide, like Hodja's friend, which village he is heading for. They must accept that they will not make any drastic changes but should be content with a limited impact on the people whose lives they touch. They must remember that hearts and minds—not just houses, machinery, and fields—are their primary concern. Theirs is a three-fold challenge: to expand people's awareness of alternatives to action; to increase their ability to make good (for them) choices among these; and to improve their competence to pursue the chosen alternative. These must be the primary goals of education in development.

In defining the characteristics of competent development workers it is helpful to consider some of the Farm School's outstanding graduates, short-course trainees, village leaders, and master farmers. Through assiduous self-improvement they have become both successful and content, and they have gained the respect of their fellow villagers.

A representative sample of graduates might include Josif, the salesman of agricultural chemicals in Crete, who instructs the farmers as he sells them chemicals; Niko, the master farmer in Euboia, who organized a youth club in his village; Haralambos, one of the top orange growers in Arta; the Theodorides brothers, one growing top quality apples and the other a skilled technician in the Cotton Research Institute; Metaxas, supervising university graduates at the Plant Breeding Institute even though he has only a Farm School diploma; and Panayiotis and his daughter Aphrodite, who together manage a pig farm with three hundred sows that produce more than five thousand piglets each year. A combination of their Farm School training and subsequent work experience prepared these people for roles as managers and leaders. But in addition to management skills they acquired other qualities that contributed to their success.

Several years ago a graduation speaker at the Farm School presented various objects to symbolize the attitudes that any village leader or development worker should possess. He first showed the students a donkey saddle, illustrative of the months and years of back-breaking work that would be required of them. Next was a Swiss pocketknife with fifteen attachments, indicating the variety of skills they would need. On his list was a plannaing book so that they would know where they were going and evaluate where they had been. As he showed the graduates a pocket calculator he pointed out that it was as important a tool as their tractor, plow, or harvester. He held up a backgammon game, emphasizing the need for recreation for both the development worker and the villagers with whom he works. An icon from the church underlined the importance of spiritual values. He held high an Olympic torch that had been used to carry the flame from Olympia in Greece to open the Olympic Games at Munich in 1972, symbolizing the significance of dreams, ideals, and ambitions.

A ROUGH ROAD AHEAD

But development work requires a lot more than the qualities referred to by the graduation speaker. It is an arduous task, and the worker must be prepared for the challenges, difficulties, and disappointments that come when adapting to a strange environment. Farm School staff members deliberated on what a development worker might anticipate, based on their own experiences and those of a large number of people who have worked in Greece. These expectations are far from universal and affect each individual quite differently. Knowing that there might be a problem does not eliminate it, but it makes it much easier to resolve.

Most development workers set out confident about what they intend to teach. It takes them some time to discover how much they have to learn, and until they understand this, their lives can be very frustrating. Young people in particular are eager to mold the world rather than change themselves. The sooner they realize that development is a two-way process the happier and more successful they will be in their work.

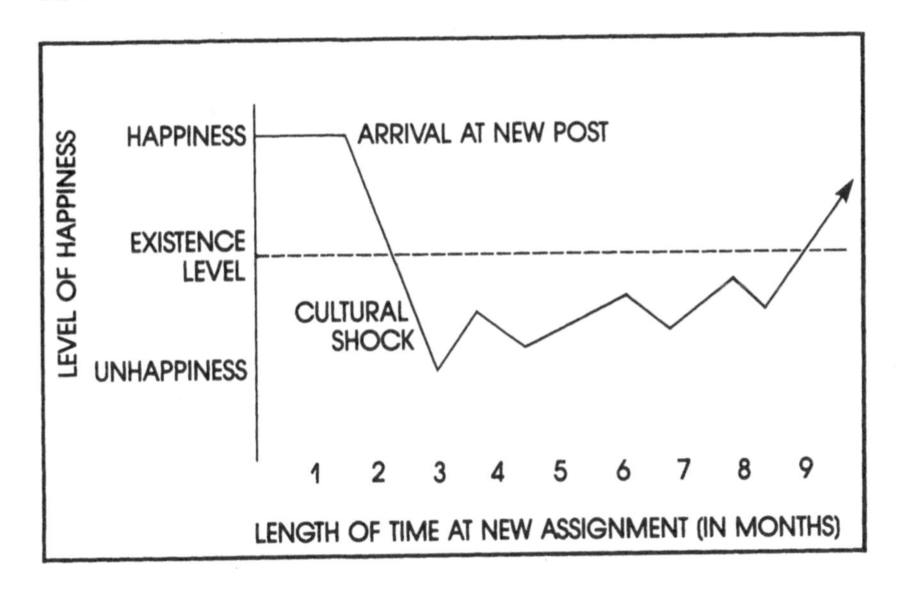

Theo Litsas is mentioned many times in this book. He was closely associated with two Farm School directors, and yet in many ways they worked for and learned from him. The younger director at first had difficulty accepting suggestions from a Greek associate who was his assistant, but Litsas had a way of teaching with a Hodja story and a smile that made his ideas more palatable.

Because their expectations are unrealistic, some people fail to anticipate the frustration associated with a new job, especially in a different culture. They usually experience depression, unhappiness, and a sense of failure before seeing any signs of success. The expatriate wife of a Farm School staff member who came to work in Greece had listened to her husband speak of the blue skies, enchanting seas, and friendly people. She became somewhat depressed after her arrival thinking that she had been deceived. In time she realized that the Greek people are like most others; that the skies are grey as well as blue, and that the sea can be both rough and calm. The Farm School uses a "quotient of happiness" line on a cultural shock graph to help expatriate staff anticipate the depression and learn to smile at themselves rather than aggravate their unhappiness. Although development workers arrive at their new posts filled with excitement about starting a new job, meeting new people, and anticipating new experiences, invariably depression sets in. The sequence of euphoria, shock, and recovery is almost unavoidable, although fortunately most can anticipate the satisfactions of acculturation in the end.

Studies of human potential in recent years have demonstrated that few human beings use more than ten to fifteen percent of their latent abilities. Living and working in a developing society demand far more than workers ever realized they could give. Like the Kazantzakis phrase quoted in the preface they must learn to say, "Lord . . . go ahead and stretch me even if I do break."

Combined with the overbearing demands on the development worker is the constant feeling of frustration growing out of the failure to attain measurable goals. Development work can be compared to the experience of the young Buddhist monk who told his master that he was meditating to become a Buddha. Without comment, the master started polishing a tile, insisting to the student that he would transform it into a mirror. When the student protested, the master replied that he had as much chance of turning the tile into a mirror as the student had of becoming a Buddha. Surely it is improbable that the development worker will accomplish a fraction of what he dreamed of in the beginning, regardless of how far he is willing to be stretched.

No matter how carefully new programs or activities are planned, many fail to materialize. Some programs initiated at the Farm School more than twenty years ago are still not fully implemented, whereas others were found to be impractical and were discontinued. The Greek villager says, "Things never happen as you want them to, but neither as you fear that they might." If development workers accept that some programs may not work out as they anticipated, they will find it easier to adjust to changing circumstances.

Getting people to change ingrained habits is a far more complex process than teaching them to manufacture automobiles, build houses, or repair television sets. There is a story of three Boy Scouts who rushed to report to their leader that they had done their good deed for the day by helping a lady to cross the street. When the Scout leader asked why it took three of them they replied, "Because she didn't want to." The erroneous assumption that all people want to cross the street in the way that the development worker thinks they should is difficult to overcome.

At the heart of most successful technical assistance programs have been a few individuals who are able to see life as a game that does not necessarily have to be won to be enjoyed.[4] A sense of humor helps those who become frustrated to smile at themselves and those around them and to accept adversity with amusement rather than with anger. It teaches the visitor to emulate the local peasants who face seemingly insoluble problems with a shrug and a smile.

A woven wall-hanging in the office of the Farm School's director quotes a time-honored Greek peasant maxim: "You have to wet your bottom to catch a fish." Over the years the most unhappy expatriates at the school have been those who had the least to do. As they became more frustrated they seemed to seek new excuses to stay home and lick their wounds, which only aggravated the situation. The best cure for such discontent has always been hard work, which invariably changes their outlook and boosts their morale.

Wylie Sheldon, a young volunteer from Princeton University, spent a year at the school. For the first few months everything seemed to go wrong for him: The students misunderstood his warmth and took advantage of him in class, and few of the staff members were willing to accept his innovative proposals. Although he often wondered why he had come to Greece, he did not give up. He devoted himself totally to his work, spending long hours preparing for classes, athletic activities, and recreation programs. He also made an effort to get close to each student—to know, understand, and help him. On his last night at the school when the students had a farewell party for him he presented them with his most prized possession—his guitar—as a farewell gesture. Spontaneously they raised Wylie on their backs, guitar in hand, and carried him round the campus singing the songs they had learned together. For him the year was filled with personal discovery as well as sharing. Wylie Sheldon's experience is reminiscent of a saying introduced to the Farm School by its most faithful volunteer, who has served it for more than thirty-five years:

> I sought my soul, my soul I could not see.
> I sought my God, my God eluded me.
> I sought my brother, and there I found all three.

Soul, God, brother—these are the elements that development workers like Sheldon seek. They will eventually discover all three if they stay long enough, but they should not anticipate much gratitude, especially in the beginning.

A PERSONAL CHALLENGE

Rural development is a building process demanding many types of workers using various materials applied in different ways. But at the center of the process are the individuals: extension agents, home economists, community development workers, local leaders, housewives, and farmers. They must cooperate to implement a carefully designed development plan, just as the carpenter, mason, plumber, and electrician work together on a building, following the architect's blueprint.

In Greece the story is told of a stranger who discovered three masons at work in a distant mountain village. When he asked them what they were doing, the first said he was constructing a wall. The second told him he was earning his living. The third reflected a moment and replied, "Oh, we're building the most beautiful church in the world." The village church is more than just the structure and the priest. If it truly exists it must be in the hearts of the people. Helping them to want to build the most beautiful church is one of the most important aspects of development.

Although government inputs for infrastructure, institutional support, extension, and marketing assistance play a significant role in development, in the long run the process depends on the efforts of the peasant family—the man and woman in the village, assisted by the development worker. Together they must solve their own problems and those of the community. The challenge in Third World countries is to help the almost eight hundred million farm workers in rural areas to cultivate the necessary skills, management, and leadership. The great majority of these people will be trained through a variety of extension, home economics, and community development programs, but experience in Greece makes it clear that those who are to become leaders in development need the more intensive instruction provided by agricultural and home management schools and short-course training centers.

Each year Farm School graduation culminates in a candlelighting ceremony comparable to that of the Easter Resurrection Service of the Greek Orthodox Church. Just before midnight the lights are extinguished in the church so that all is in darkness except for one small lamp over the holy table. At the stroke of midnight the priest takes light from the lamp and shares it with the elders, who in turn light the candles that the members of the congregation have brought with them. As the candlelight spreads in waves it illumines the whole church, inspiring a spirit of warmth and unity among the parishioners, who take their lighted candles home at the end of the service. The graduation ceremony is used to symbolize the tools of progress that Farm School graduates have carried back to their villages for more than three-quarters of a century. Training managers for rural development in this or any other training center is the process of lighting a flame and ensuring that it continues to spread.

Increasing numbers of peasant families who have attended training centers throughout Greece have become effective managers and raised the country's production level and standard of living to unbelievable heights, working as farmers, homemakers or technicians servicing agriculture. Just as the priest and his church are vehicles for transmitting the light and inspiring the members of the parish, so training centers have cultivated a climate in which the peasants themselves wanted to learn to become capable managers. Without the help they received from these centers it would have been difficult for them to understand and accelerate the process that has enriched their lives. Although training centers are an essential link, ultimately the vanishing peasants become the master farmers, the nikokiris, the dynamic leaders in their communities—using their heads, hands, *and* hearts—who have turned their dreams into reality.

Notes

Preface

1. *Encyclopedia Britannica*, vol. 17 (Chicago, London, Toronto: 1955), p. 425.

2. This is a very free translation of the opening statement of Nikos Kazantzakis, *Report to Greco*, translated by P. A. Bien (London: Faber and Faber, 1973), p. 16.

Acknowledgments

1. "Γηράσκω δ' αιει πολλα διδασκόμενοζ." Quoted by Plato, who attributed this quotation to Solon. From "Ερασται" (Lovers), 133-C-6, in *Platonis Opera*, vol. 2, Tetralogias III–IV, OXIONI, edited and translated by John Burnet (Cambridge: Cambridge University Press, 1901). This phrase has also been attributed to Heraclitus.

Chapter 1

1. Julius K. Nyerere, *Ujamaa: Essays on Socialism* (Tanzania: Oxford University Press, 1968).

2. John Henry House described his thoughts about education in the period between 1880 and 1883 (American Farm School Archives).

3. Stephen D. Salamone, "The *Nikokiris* and the *Nikokira:* Sex Roles in a Changing Socio-Economic System" (unpublished manuscript).

Chapter 2

1. "Statistiki Fisikis Kinisis Plithismou etous 1980" (Statistics of national movement of population for year 1980) (Athens: Ministry of National Economy, National Statistical Service, 1983), p. 46. Projections are based on figures for 1928 and 1970.

2. "Statistiki Epetiris tis Ellados etous 1981" (Statistical register of Greece for year 1981) (Athens: Ministry of National Economy, National Statistical Service, 1983), p. 41. Projections are based on figures for 1884 and 1921.

3. Estimates were provided by agricultural economic specialists Stephanos Lazaredes and Nicolaos Galanis, the Inspectorate of Agriculture of the Ministry of Northern Greece, during a personal interview in October 1983.

4. William Papas, *Instant Greek* (Athens: Self-published by William Papas, 1972, distributed by A. Samouches).

5. J. K. Campbell, *Honour, Family and Patronage* (Oxford: Oxford University Press, 1964). This is an excellent book on the Sarakatsani, the nomadic shepherds of Greece.

6. W. H. McNeill, *The Metamorphosis of Greece Since World War II* (Chicago: University of Chicago Press, 1978), and *Greece: American Aid in Action 1947–56* (New York: Twentieth Century Fund). McNeill's next report should provide interesting insights on the impact of Greece's entry into the Common Market in 1980.

7. McNeill, *Metamorphosis*, pp. 138–205.

8. McNeill, *Metamorphosis*, p. 11.

Chapter 3

1. Brenda L. Marder, *Stewards of the Land: The American Farm School and Modern Greece* (Boulder, Colo.: East European Quarterly, 1979, distributed by Columbia University Press, New York).

2. Notes from the diary of Robert J. Rawson, who visited the Farm School in 1920. Original notes are in the Farm School archives.

3. Irwin T. Sanders, Raymond W. Miller, and Robert W. Miller, "Report on the American Farm School of Thessaloniki, Greece," report prepared for the Board of Trustees, July 30, 1955.

Chapter 4

1. Kenneth Blanchard and Spencer Johnson, *The One Minute Manager* (New York: William Morrow, 1982).

2. CIAT, Centro Internacional de Agricultura Tropical, Cali, Colombia. Publicity pamphlet (undated), distributed by the center.

3. List was provided by Ann Powers during a personal interview with the author at An Grianan in Ireland, 1980.

Chapter 5

1. Irwin T. Sanders, "Future Directions for the Thessalonica Agricultural and Industrial Institute," report prepared for the Board of Trustees, October 1977, pp. vi–vii.

2. Paul H. Davis, "All the World Stands Aside," *Association of American Colleges Bulletin* 43, no. 2, May 1957.

3. Antonios Trimis, "Community Development as an Element in Area and Regional Socio-economic Growth and Development (with special reference to the Community Development Program of Thessaloniki, Greece)," Ph.D. dissertation, Montana State University, June 1967. Special credit should also be given to Lotta Hitschmanova of the Unitarian Service Committee of Canada, who provided enormous moral and financial support for the program.

4. Daniel Burnham, quotation from notes compiled by Ruth D. Wells, "Two Centuries of Burnhams, 1735–1946."

Chapter 6

1. Ernestine Friedl, *Vasilika, a Village in Modern Greece* (New York: Holt, Rinehart and Winston, 1962).

2. Clio Presvelou, *More is Not Better* (Wageningen, Holland: Department of Home Economics, 1980).

3. Nomiki Tsoukala, O *Desmos tis Agrotikis Oikiakis Oikonomias,* Master's thesis, University of Tennessee, 1977.

Chapter 7

1. Paraphrased from an unpublished circular, "Leadership," by Jerold Panas & Partners, Inc., Chicago.

2. Harvey Mindess, "A Sense of Humor," adapted from his book *Laughter and Liberation*, in *Saturday Review of Literature*, August 21, 1971.

3. Henri Nouwen, *With Open Hands* (Notre Dame, Ind.: Ave Maria, 1982).

4. The ideas in this letter are from a mimeographed statement by Dr. R. A. Hatcher of Georgia State University.

5. Brenda L. Marder, *Stewards of the Land: The American Farm School and Modern Greece* (Boulder, Colo.: East European Quarterly, 1979, distributed by Columbia University Press, New York).

6. Daniel Benor and James Q. Harrison, *Agricultural Extension: The Training and Visit System* (Washington, D.C.: World Bank, 1977).

Chapter 9

1. Paraphrased from a Rochester, N.Y., Rotary International publication, circa 1950.

2. These comments are based on a course taught by Professor Tad Hungate at Teachers College, Columbia University, New York, in 1953.

Chapter 10

1. Douglas McGregor, *The Human Side of Enterprise* (New York: McGraw-Hill, 1960).

2. This concept was explained by Ita Hartnett during a personal interview in 1980.

3. Based on an interview with Professor Johannes Dopmeyer and his staff.

Chapter 11

1. *National Agricultural Occupations Competency Study* (national study for identifying and validating essential agricultural competencies needed for entry and advancement in major agriculture and agribusiness occupations), U.S. Department of Health, Education and Welfare, Office of Education, Occupational and Adult Education Branch, Washington, D.C., May 16, 1978.

2. John R. Crunkilton and Alfred H. Krebs, *Teaching Agriculture Through Problem Solving* (Danville, Ill.: Interstate Printers and Publishers, 1982).

3. Crunkilton and Krebs, *Teaching Agriculture Through Problem Solving.*

4. Curtis R. Finch and John R. Crunkilton, *Curriculum Development in Vocational and Technical Education* (Boston: Allyn and Bacon, 1979).

5. "Aristotle on Education," extracts from the *Ethics and Politics*, translated and edited by John Burnet (Cambridge: Cambridge University Press, 1903, 1967). "The things which we are to do when we have learnt them, we learn by doing them; we become, for instance, good builders by building and good lyre-players by playing the lyre" (p. 45). "It is well worth while to note carefully all that Aristotle says as to the relation between theory and practice in education. He does not disparage theory, indeed he holds that it is absolutely impossible for us to do anything without it. What he does object to is the tendency to put theory in the place of practice" (fn. p. 52). Note also, the *Politics* of Aristotle from *The Works of Aristotle Translated into English*, vol. 10, translated by Benjamin Jowett, rev. ed. (Oxford: Clarendon Press, 1885, 1921), "Now it is clear that in education practice must be used before theory, and the body before the mind" (1338 6 5–6).

Chapter 12

1. Information in this paragraph comes from an interview with Professor Anne Van den Ban of Wageningen University.

2. These observations were made by Frank Madaski, extension district supervisor, who has headquarters at Michigan State University, Lansing, Michigan.

3. These comments are based on a personal interview with C. Presvelou, professor of home management at Wageningen University. Further details are available in her inaugural address "More is Not Better" (Wageningen, Holland: Department of Home Economics, 1980).

4. The sections that follow are based on an inspiring discussion with Ita Hartnett in County Cork in 1979. It is unfortunate that people with her insight become so involved in their work that they do not have time to write about it.

Chapter 13

1. Edward de Bono, *Teaching Thinking* (Middlesex, England: Penguin Books, 1976); see also de Bono, *Lateral Thinking: A Textbook of Creativity* (Middlesex, England: Penguin Books, 1970).

2. Theodore Schultz, "The Economics of Being Poor," Nobel lecture, Stockholm, Sweden, December 8, 1979.

3. James L. Adams, *Conceptual Blockbusting—A Guide to Better Ideas* (New York: W. W. Norton, 1974). This book is based on the author's years of teaching engineering management at Stanford University.

4. Adam Smith, *Powers of Mind* (New York: Random House, 1975).

5. Adams, *Conceptual Blockbusting.*

Chapter 14

1. W. D. Ross, *Ethica Nicomachea* (London: Oxford University Press, Humphrey Milford, 1925), Book 1.7, 1097a 39 to 1097b 7. "Now such a thing happiness, above all else, is held to be; for this we choose always for itself and never for the sake of something else, but honour, pleasure, reason and every virtue we choose indeed for themselves . . . but we choose them also for the sake of happiness, judging that by means of them we shall be happy. Happiness, on the other hand, no one chooses for the sake of these, nor, in general, for anything other than itself."

2. Theodore W. Schultz, "The Economics of Being Poor," Nobel lecture, Stockholm, Sweden, December 8, 1979.

3. The theory of Transactional Analysis is based on Eric Berne's original thesis in *Games People Play* (New York: Grove Press, 1964). It was further developed by Thomas A. Harris in *I'm OK—You're OK* (New York: Avon, 1967), and Muriel James and Dorothy Jongeward in *Born to Win* (Manila, Philippines: Addison-Wesley, 1971).

4. See note 3 for complete references.

5. Berne, *Games People Play.*

6. Harris, *I'm OK—You're OK.*

7. Alvyn Freed, *T. A. for Teens* (Sacramento, Calif.: Jalmar Press, 1976).

8. Muriel James and Louis M. Savary, *The Power at the Bottom of the Well* (New York: Harper and Row, 1974).

9. Henri J. M. Nouwen, *Reaching Out—The Three Movements of the Spiritual Life* (New York: Doubleday, 1975).

Chapter 15

1. *The Random House Dictionary* (New York: Random House, 1967), p. 900.

2. Ovid, *Metamorphoses,* translated by Rolfe Humphries (Bloomington, Ind.: Indiana University Press, 1955).

3. Franz Kafka, *The Metamorphosis,* translated and edited by Stanley Corngold (New York: Bantam Books, 1972), pp. 1–58.

4. Harvey Mindess, "A Sense of Humor," *Saturday Review of Literature,* August 21, 1971, p. 138.

Books About Nasredin Hodja

Mehmet Ali Birand. *Stories of Hodja.* Istanbul.

Downing, Charles. *Tales of the Hodja.* London: Oxford University Press, 1964.

Shah, Idries. *The Exploits of the Incredible Mulla Nasrudin.* New York: Simon and Schuster, 1966.

———. *The Pleasantries of the Incredible Mulla Nasrudin*. London: Jonathan Cape, 1968.

Shah, Idries, and Richard Williams. *Once the Mullah—The Subtleties of the Inimitable Mulla Nasrudin*. London: Jonathan Cape, 1973.

Soloviof, Leonida. O *Nastredin Xotzas*. Athens: Dorikos, 1980.

Yagan, Turgay. *Stories of the Hodja*. Istanbul: Turgay Yagan.

Appendix A

New technology and practices are continually being introduced, at the Farm School usually for the first time, in Greece. The following list has been compiled from the Farm School's records dating back to 1912.

1912	Introduction of California vines resistant to Filoxera infecting local vines
	Silo for production of silage
1914	Reaper and binder
1918	Chick incubators
1921	Introduction of Sudan grass as forage crop
1922	Manure spreader
	Silage cutter and blower
	Two-horse riding cultivator
	Introduction of Rhode Island Reds, Anconas and Leghorn chickens
	Introduction of Jersey cows and Southdown sheep
1926	Babcock tester for milk and cream
	Introduction of Gambuzi fish for control of malaria
1932	Large black pigs
1933	Guernsey cows
	Introduction of improved Babcock milk fat tester
1934	Hay loader

1935	Milk pasteurizer and bottling plant
	Introduction of improved broilers
	Stationary silage cutter and blower
	Pull-type grain combine (International Harvester)
1936	Hay mower and side delivery rake
	Introduction of sorghum grown for animal feed
1946	Introduction of kudzu as forage crop
1950	Self-propelled combine
1951	Pull-type P.T.O. (power takeoff) manure spreader
1956	Small hand-leveling machine suitable for farm use (land-level Eversman)
1957	Introduction of sweet corn, broccoli, and Brussels sprouts
1958	Mechanical fertilizer distributor (New Holland 410)
	Trench silos
1962	Cotton picker by vacuum
	Corn picker (New Idea N 310)
1963	Mower conditioner (Swather)
	Open-shed, free-housing (not tied) system for cattle
1964	Subsoiler for breaking under surface hardpan (Killifer John Deere No. 20)
1965	Onion seed
	Large-scale calf importation, by plane (with the Ministry of Agriculture)
1966	New Holland Forage Harvester
	Hay harvester (rotary cutter) for cutting crop residue (Cyro 80)
	Introduction of wheat seed from Mexico
	Automatic milking for large herd
	Dairy cows artificially inseminated with frozen semen
1967	Automatic milking for family-size dairy herd
1968	Introduction of Cryovac system for packing poultry
1969	Introduction of hybrid imported turkeys
1971	Mower conditioner to speed up hay drying (New Holland 469 Haybine)
1975	Slotted floor barn for heifers
	Preparation of skills charts for competency training
1976	Preparation of 375 basic competencies for Farm School students
	Pull-type bale wagon for single operator, collection and transportation (New Holland 1006)
1977	Sow artificially inseminated with imported frozen semen
	Pull-type automatic baler loaders
1978	Rear-mounted subsoiler (Killifer McCormick)
	Irrigation gun
	Coeducational scholi/lyceum program
	Introduction of teaching package concept to Farm School students
1980	Introduction of teaching package concept for short-course centers
1981	Turbine combine without straw rack and return pan for low vibration and speedier operation (New Holland)
	Short courses for farmers under Common Market auspices

1982 Launch site for new fuel-efficient models of Ford tractors
Introduction of exotic trees and shrubs from abroad
Installation of mist propagation

1983 Creation of Alternate Energy Center with following units:
Parabolic collector for generating intense heat at focal point
Hydronic Thermosyphon Collector for hot water
Solar oven
Solar food dryer
Establishment of a student computer center
Establishment of a periodicals distribution center for graduates
Establishment of an audiovisual center to prepare new training materials
Biogas plant for anaerobic digestion and stabilization of animal waste and methane recovery

Appendix B

THE AMERICAN FARM SCHOOL
Thessaloniki, Greece
Teacher's Instruction for Use of Teaching Package

The teaching package is a guide for the teacher's use in teaching a lesson.

The teaching package includes the following.

I. Introduction

A. Objectives
This is a brief description of what the trainee is expected to learn. It may be information that the trainee needs to know or skills and procedures that the trainee needs to develop.

B. Rationale
This is an explanation describing WHY the lesson is important to the trainee. The rationale includes key questions and problems of personal importance to the trainee. The trainee must be convinced that the information in the lesson is important to him or he will have no interest in it.

C. Summary
This is a summary of the material in the lesson that the teacher is presenting.

II. Instructional Activities

A. Presentation of the Subject

The teacher will present the lesson in the following manner:

1. Describe the objectives of the lesson.
2. Explain the rationale of why the lesson is important.
3. Give a brief summary of the material to be presented in the lesson.

B. Additional Learning Activities

Any learning activity that will assist the trainee in achieving the objectives is appropriate. Each of the following activities is effective:

1. Using transparencies
2. Giving a demonstration
3. Showing slides, filmstrips, or movies
4. Showing charts
5. Lecturing
6. Writing on the blackboard
7. Using handout material
8. Adding resource personnel

C. Practical Exercises for Trainees

It is important to involve the trainees in practical exercises relating to their problems and the objectives of the lesson. After a demonstration or explanation, there may be appropriate practical exercises for the trainees in order to reinforce the learning process. This is also a good way for the teacher to determine if the trainees have learned the lesson or developed the skill as described in the objectives.

D. Evaluation

1. By teacher—At this point in the lesson, the teacher should ask the trainees questions to determine if the objectives of the lesson have been achieved. The teacher should also ask the trainees questions during the lesson to determine if they comprehend the material that is being presented.
2. Self-evaluation—Near the end of the lesson, a short written quiz should be completed by the trainees to determine how well they have comprehended the lesson. Four questions should be asked with multiple choice or true and false answers. The correct answers should be given on the following page with an explanation of why the answers are correct. This quiz is for the trainees' use only and is not handed in to the teacher.

E. Review of Lesson

At the end of the lesson, the teacher should review briefly the materials that have been presented. The review should include the objectives, rationale, and key questions and answers. The teacher should also ask if the trainees have any questions in order to clarify any confusing points.

III. Teaching Aids

A. Bibliography

This is a list of reference material, books, magazines, reports, and bulletins that were used to prepare the teaching package. Teachers and trainees are encouraged to read these materials, if available, in order to become more familiar with the subject.

B. Audiovisual Materials

This is a list of audiovisual equipment and materials needed to teach the lesson. It may include a slide projector and slides, an overhead projector and transparencies, or a movie projector and movie film. It will also include any printed materials that will be distributed to the participants. It will also include any equipment needed for a demonstration.

Note: The major items listed are in chronological order and should be presented in the lesson in the prescribed manner; first A, then B, then C, and so on.

Teachers are encouraged to add additional examples and questions in order to improve the effectiveness of instruction. It is also very important for the teacher to involve the trainees in discussion. Trainees must have the opportunity to relate their own experiences and to ask their own questions. They must have a sense of participation, or they will ignore the teacher and lesson.

Copies of some of the materials of the teaching package should be available for distribution to the trainees for additional reference after returning to their farms. They should be distributed at the time of the self-evaluation quiz and NOT before; otherwise trainees will be too busy reading the materials and will not pay attention to the teacher and the lesson.

Reminders:

1. The teaching package should be used as a guide and teachers are encourged to supplement the materials with personal examples and questions in order to improve relevancy and effectiveness of instruction.

2. The trainees' participation and involvement throughout the lesson are mandatory for effective learning.

3. At the end of the teaching package, the teacher should determine if the trainees have achieved the objectives.

Index